ARKANA

JOY

Dr. Alexander Lowen is the creator of bioenergetics, a revolutionary method of psychotherapy designed to restore the body to its natural freedom and spontaneity through a regimen of exercise. The foremost exponent of this method of incorporating direct work on the body with the psychoanalytic process, Dr. Lowen practices psychiatry in New York and Connecticut and is the executive director of the Institute of Bioenergetic Analysis. He and his wife live in New Canaan, Connecticut. Dr. Lowen's other titles include *Depression and the Body*, *Bioenergetics*, *Pleasure* and *Love, Sex and Your Heart*, all of which are available from Arkana.

PENGUIN

ARKANA

JOY

The Surrender to the Body
and to Life

Alexander Lowen, M. D.

ARKANA

ARKANA

Published by the Penguin Group
Penguin Books USA Inc., 375 Hudson Street, New York, New York 10014, U.S.A.
Penguin Books Ltd, 27 Wrights Lane, London W8 5TZ, England
Penguin Books Australia Ltd, Ringwood, Victoria, Australia
Penguin Books Canada Ltd, 10 Alcorn Avenue, Toronto, Ontario, Canada M4V 3B2
Penguin Books (N.Z.) Ltd, 182–190 Wairau Road, Auckland 10, New Zealand

Penguin Books Ltd, Registered Offices: Harmondsworth, Middlesex, England

First published in Arkana 1995

3 5 7 9 10 8 6 4 2

LIBRARY OF CONGRESS CATALOGING IN PUBLICATION DATA
Lowen, Alexander.
Joy: the surrender to the body and to life/Alexander Lowen.
p. cm.
ISBN 0 14 01.9493 2 (pbk.)
1. Bioenergetic psychotherapy. I. Title.
RC489.B5L685 1995
616.89'14—dc20 95–17549

Printed in the United States of America
Set in Granjon
Designed by Brian Mulligan
Illustrations by Kris Tobiasson

This book is dedicated to my wife,
Rowfreta Leslie Lowen,
a truly religious person

CONTENTS

It is forty-eight years since I saw my first client in therapy. I had just completed my analysis with Wilhelm Reich. Reich's work was becoming known, resulting in a demand for his kind of therapy. As there were very few individuals trained in his approach, I was sought out despite the fact that I was not a physician at the time. As a beginner I charged my first patient two dollars an hour, which was a small fee even then. But as I look back to that early experience, I question if I was worth even that small sum. I had no idea of the depth and severity of the disturbance that afflicts so many people in our culture—the depression, the anxiety, the insecurity and the lack of love and joy in life.

After working with people for almost half a century, during which time I have written eleven books, I believe I have gained an understanding of the human problem and have formulated the principles of an effective therapeutic approach which I call Bioenergetic Analysis. This book will describe the process of that therapy and illustrate its application through the case histories of my clients. Let me say that it is not a quick and easy cure, though it is effective. However, its effectiveness depends upon the experience and self-understanding of the therapist. Since the problems that people struggle with have been structured in their personalities for many years, it is unrealistic to expect any quick or easy cure. True miracles rarely happen. The one miracle that occurs regularly is the miracle of the creation of new life. To that miracle, this book is dedicated.

The underlying theory of Bioenergetic Analysis is the functional identity and antithesis of mind and body—of psychological and physical processes. Functional refers to the fact that body and mind act as a unit in running the body, and are a unit on the deep level of the energetic processes. The antithesis is reflected in the fact that on the surface the mind can influence the body and the body, of course, affects thinking and mental processes.

Bioenergetic Analysis is based on the concept that a person is a unitary being and that what happens in the mind must also be happening in the body. Thus, if a person is depressed with thoughts of despair, helplessness and failure, his body will manifest a similar depressed attitude, evident in decreased impulse formation, reduced mobility and restricted breathing. All bodily functions will be depressed, including metabolism, resulting in lowered energy production.

Of course, the mind can influence the body just as the body affects the mind. It is possible in some cases to improve one's bodily functioning through a change in one's mental attitude, but any change so induced will be temporary unless the underlying bodily processes are significantly changed. On the other hand, directly improving bodily functions such as breathing, moving, feeling and self-expression has an immediate and lasting effect on one's mental attitude. In the final analysis, increasing a person's energy is the fundamental change which the therapeutic

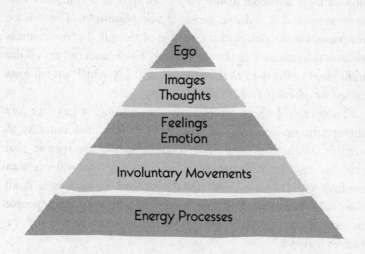

HIERARCHY OF PERSONALITY FUNCTIONS

process must produce if it is to reach its goal of freeing an individual from the restrictions of his past and the inhibitions of the present.

The above diagram shows the hierarchy of personality functions as a pyramid with the ego at the top. These functions are interrelated and dependent on each other, and all rest on a base representing energy production and use.

The aim of therapy is to help an individual recover the full potential of his being. All individuals who come to therapy have been greatly diminished in their capacity to live and experience the fullness of life by the traumas of their childhood. This is the basic disturbance in their personality which underlies the symptoms they present. While the symptoms denote how the individual has been crippled by his upbringing, the bottom line is the loss of part of the self. All patients suffer from some limitation in their selfhood: a limited self-awareness, a restricted self-expression and a reduced sense of self-possession. These basic functions are the pillars of the temple of the self. Their weakness creates an insecurity in the personality which undermines all the individual's efforts to find the peace and joy which give life its fullest satisfaction and deepest meaning.

To overcome these limitations is an ambitious goal for any therapeutic undertaking, and, as I said before, it is not easy to achieve. Without a clear understanding of the therapeutic goal one can get lost in the maze of conflicts and ambivalence that confuse and frustrate most therapeutic endeavors. But it is an essential undertaking that can substantially help the many people in our culture for whom life is a struggle to survive and joy is a rare experience.

ACKNOWLEDGMENTS

I would like to thank David Randolph, who supplied the references about joy in the Bible and to Michael Conant and Leslie Case, who read the manuscript and offered some suggestions.

JOY

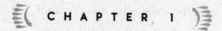

CHAPTER 1

JOY

The Freedom from Guilt

When I work with my patients, most of them will leave the session feeling good. Some will leave feeling joyful. But these good feelings generally do not last. They result from the experience during the session of breaking free from some restricting tension, of feeling more alive and of understanding one's self more deeply. They do not last because the breakthrough was achieved with my help and the patients are not able to maintain their openness and freedom alone. But each breakthrough of feeling, each release of tension, is a step toward the recovery of the self even if one cannot fully hold the gain. These are also temporary gains because the patient, as he goes deeper into himself physically and psychologically in his search, will encounter more frightening memories and feelings from an earlier period of his childhood—feelings which have been more deeply suppressed in the interest of survival. But as one goes deeper into the self one also gains courage to deal with these early fears and traumas in a mature way, that is, without denial and suppression. Somewhere deep inside each of us is the child who was innocent and free and who knew that the gift of life was the gift of joy.

Young children are generally open to feelings of joy. They are known to jump for joy, literally. Young animals do the same, kicking up their heels and running about in a joyful abandon to life. It is very rare to see a mature or older person feel and act that way. Dancing may be the closest they come to it, which is why dancing is the most natural activity at joyful occasions. Children, however, do not need a special occasion to be joyful. Allow them to be free in the company of other children and joyful activity will soon appear. I remember when I was four or five years old, and was out on the street with several other children as it began to snow heavily. We were excited and we began to dance around a light pole chanting, "It's snowing, it's snowing, a little boy is growing." I have always remembered the joy I felt on that occasion. Children often feel a sense of joy when they receive a gift of a much-desired object, which leads them to jump and squeal with delight. Adults are more restrained than children in their expression of all feelings, which limits the intensity of their good feelings. In addition, they are burdened by cares and responsibilities and beset by guilt, which dampens their excitement so that joy is rarely experienced.

I have known joy on some very ordinary occasions. Walking along a country road some time ago, I felt my spirit soar. It was a familiar road, with no special quality, but as I took a step and felt my feet press on the ground, I sensed a current go through my body, which seemed to grow two inches taller. Something released inside me and I felt joyful. Some trace of that feeling has persisted in me since that day, and although there have been some painful and disturbing episodes in my life since then, I can sense a good feeling in my body most of the time. I attribute this good feeling to the therapy I have had, starting in 1942, and to the work I have continuously done with myself over the years.

The therapy enabled me to make contact with the child in me, who knew some joy despite a basically unhappy childhood, and to integrate into my adult life those qualities of childhood which make joy possible.

Childhood—assuming that it is a healthy, normal childhood —is characterized by the two qualities which lead to joy: freedom and innocence. The importance of freedom to the feeling of joy needs little explanation. It is difficult to imagine anyone feeling joyous whose movements are restricted by some outside force. When I was small the most dreaded punishment my mother could inflict upon me was to make me stay indoors on a day when other boys were out playing. One of the reasons I, like so many other children, was anxious to grow up was to gain my freedom. As I reached maturity I became free from parental control. In this culture, freedom means the right to pursue one's own happiness or joy. Unfortunately, external freedom is not enough. One must also have internal freedom, namely, the freedom to express one's feelings openly. I didn't have that freedom, nor do many people in our culture. Our behavior and expressions are controlled by a superego, with its lists of Do's and Don'ts and the power to punish if one violates its commandments. The superego is the internalization of the "dictatorial" parent. It functions, however, below the level of consciousness so that we are not aware that the limitations it imposes upon our feelings and actions are not the result of our free will. Dethroning the superego and restoring an individual's freedom of expression does not turn him into an uncivilized being; rather, it is a condition that allows him to be a responsible member of society, a truly moral person. Only a free person is respectful of the rights and freedom of others.

Nevertheless, we must recognize that social living requires some restrictions upon our individual behavior in the interest of group harmony. All human societies regulate the social behavior of its members but those regulations judge actions, not feelings. An individual may be judged guilty if he violates the accepted social code of behavior but he is condemned for the transgression alone and not for a feeling or desire. Civilized societies based on power extend the concept of guilt to include thoughts and feelings in addition to actions.

This change is exemplified in the biblical story of Adam and Eve. The Bible details how, by eating the fruit of the tree of knowledge, they lost their innocence and joyfulness. Prior to eating the forbidden fruit from the tree of knowledge they lived in a state of bliss in the Garden of Eden, the original paradise, as animals among the other animals, following the natural instincts of their bodies. Once they ate the forbidden apple, they knew right from wrong, good from evil. Their eyes were opened, and they saw that they were naked. They covered themselves because they were ashamed and they hid from God because they felt guilty. No other animal knows right from wrong, has a sense of shame or feels guilty. No other animal judges its feelings, its thoughts and its actions. No other animal judges itself. No other animal can conceive that it is "good" or "bad." No other animal has a superego or is self-conscious, unless it is a dog who lives in a dependent relationship in the home of its owners, very much as children do.

We train our dogs to observe certain patterns of behavior which we consider right or good and punish or humiliate them when they disobey. A dog that fails to obey is often called "bad dog," and most dogs do learn how to behave to please their masters. Teaching a dog or a child how to behave in a civilized setting is necessary for social living, and both a dog and a child will naturally try to conform to what is expected if the expected behavior does not violate their integrity. Too often, however, that integrity *is* violated, causing the animal or the child to resist, leading to a power struggle it cannot win. It ultimately submits to the violation which, in effect, breaks its spirit. One can observe that break in a dog who cowers with its tail between its legs before an owner, but one can also see the break in a child whose eyes become dull, whose body stiffens and whose manner is submissive. Such children grow into neurotic adults who may know how to win but do not know how to be joyful.

People who come to therapy, no matter how successful they may be in their careers, are individuals whose spirit has been broken to the degree that joy is an alien feeling. The particular symptoms they present are merely the outward manifestations of their distress. Some have been broken to where they are dysfunctional, while others do manage to function in society. It is a mistake to assume that because one doesn't go to therapy, or doesn't believe it is needed, one is healthy. I started my therapy with Reich under the illusion that I was okay, but it did not take me long to discover that I was in fact frightened, insecure and physically tense in my body. In a previous book, *Bioenergetics*, I related some of my experiences in this therapy, which shocked me with the realization of the extent of my neurosis, but showed me the way to recover my integrity and gave me the courage to follow my chosen path.

That path was the surrender to the body. What I had to surrender was my identification with my ego in favor of an identification with my body and its feelings. On an ego level I saw myself as bright, intelligent and superior. I believed I could accomplish much, but what I could accomplish I did not know. I desired to be famous. I was driven by an inordinate ambition instilled in me by my mother to make up for the lack of ambition in my father, but fortunately I had enough support from him to prevent my mother from dominating me. Surrender to the body would involve giving up this inflated ego image which covered and compensated for underlying feelings of inferiority, shame and guilt. If I accepted these feelings, I would feel terribly humiliated—something I was unconsciously trying to avoid. The surrender to the body entails the surrender to sexuality, which I sensed was at the root of my deepest fears of rejection and humiliation. And yet it was the lure of the joy and ecstasy of sex that drew me to Reich and led me into therapy with him.

On a conscious level I didn't feel guilty about my sexuality. As a sophisticated, modern adult I could accept sexuality as natural and positive. But on a body level I felt driven by a desire which found no deep satisfaction. I was a typical narcissistic individual, who seems free in his sexual behavior but whose freedom is external not internal, a freedom to act but not to feel. I would have denied any feeling of guilt about sexuality, but I could not surrender fully to any woman and I could not allow sexual excitement to overwhelm me in the sexual act. Like most individuals in our culture, my pelvis was locked by chronic muscular tensions and unable to move freely and spontaneously at the climax of the sexual act. When I finally let go of those tensions in the course of my therapy with Reich and my pelvis moved freely and spontaneously in harmony with my breathing, I felt a sense of joy such as one might feel on being released from prison.

Chronic muscular tension in different parts of the body constitutes the prison that prevents the free expression of an individual's spirit. One finds these tensions in the jaw, the neck, the shoulders, the chest, the upper and lower back and the legs. They manifest the inhibition of impulses which the person dare not express for fear of punishment, verbal or physical. The threat of rejection or the withdrawal of love by a parent is life-threatening to a young child and often evokes more fear than physical punishment. The child who lives in fear is tense, anxious and contracted. It is a painful state and the child will deaden himself to not feel the pain or the fear. Deadening the body eliminates the pain and the fear because the "dangerous" impulses are effectively imprisoned. Survival thus seems assured, but the repression becomes a mode of life for the individual. Pleasure is subordinated to survival, and the ego, which originally served the body in its desire for pleasure, now controls the body in the interest of security. A split develops between the ego and the body which is controlled by a band of tension at the base of the skull, breaking the energetic connection between the head and the body—between thinking and feeling.

Ensuring survival is one of the ego's functions as the representative of the instinct for self-preservation. It does this by virtue of its ability to coordinate the body's response to external reality. Through its control of the voluntary musculature, it takes command of all bodily functions which could interfere with survival. But like the general who becomes a dictator after tasting the power of command, the ego is reluctant to surrender its hegemony. Despite the fact that the danger is past—that the frightened child is now an independent adult—the ego cannot allow itself to accept the new reality and surrender control. It has now become a superego which must maintain control out of fear that anarchy would result if it abandoned its position. I have known many patients who as independent adults were still afraid of their parents, even afraid to speak openly to them. Face-to-face with their parents they cower like a frightened dog. When as a result of therapy they gain the courage to speak freely to the parent, they are amazed that this person whom they saw as so threatening is no longer the monster they feared.

The difference between the ego and the superego is that the former has the ability to surrender control when the situation allows. That isn't true of superego control. Very few people, almost none, can consciously relax their tight jaws, their tense neck muscles, their contracted back muscles or their stiff legs. In most cases they are not even aware of the tension and the unconscious control it represents. Many people feel the tension in their bodies because of the pain it creates, but they have no idea that the tension and the pain are the result of how they function or how they hold themselves. Some regard their rigidity as a sign of strength, as proof that they can stand up to adversity, that they will not break or yield under stress, that they can tolerate discomfort, even distress. I believe we have become a nation of survivors so frightened of illness and death that we are unable to live as free people.

This fear of surrendering egocontrol is the basic cause of our misery and discontent. Yet most people are not aware of how frightened they are. Every chronically tense muscle in the body is a frightened muscle, or it would not hold so tenaciously against the flow of feeling and life. It is also an angry muscle, since anger is the natural reaction to forced restraint and the denial of freedom. And there is sadness at losing the potential for a state of pleasurable excitation that would set the blood coursing, the body vibrating and the waves of excitation flowing through the body. Such a state of aliveness is the physical basis for the experience of joy, as many religious people know. It is in pursuit of that state of excitement that the Shakers shake, the holy rollers roll and the whirling dervishes dance until they reach ecstasy.

Joy is a religious experience. In religion it is associated with surrender to God and the acceptance of his grace. At the heart of biblical belief is the injunction, "You shall rejoice before the Lord your God." This statement, found in Deuteronomy 2:13, is Moses' counsel to the children of Israel after their deliverance from captivity in Egypt. The Hebrew word for joy is *gool*. Its primary meaning is to spin around under the influence of violent emotion. This word, which the Psalmist used to describe God, pictures him as whirling with sublime delight.

In the New Testament (John 15:11) Jesus said that he taught so that his followers may be joyful. In Hebrews 12:2 he also said, "These things I have given you, that my joy may be in you and that your joy may be full." Christianity teaches that to be at one with God, the Father, is to experience joy.

Another view of joy is given in Schiller's poem "Ode to Joy,"[1] in which joy is described as fashioned from celestial flame with the power to entice the blossom from the bud, draw sun from the sky and "set spheres through boundless ether spinning."

These images suggest that the God in heaven can be identified with those cosmic forces which create the stars. Of these stars, the most important one for life on earth is our Sun. It is the celestial flame, the spinning sphere whose rays make the earth fertile. When it shines, it lights up and warms the earth, setting in motion the dance of life. For many creatures waking to a bright, sunny day fills them with joy. The human creature is particularly sensitive to this celestial flame. It is not surprising therefore that the ancient Egyptians worshipped the sun as a God.

Rabindranath Tagore, the Indian scholar and sage, also speaks of joy in terms of natural processes. "Compulsion is not the final appeal to man but joy is, and joy is everywhere. It is in the earth's green covering of grass, in the blue serenity of the sky, in the restless exuberance of spring, in the silent abstinence of winter, in the living flesh that animates our bodily frame, in the perfect poise of the human figure—noble and upright—in living, in the exercise of all our powers. . . . Only he has attained the final truth who knows that the whole world is a creation of joy."[2]

But, one could ask, what about sorrow? We all know that there is sorrow in life. It comes to each of us in the loss of someone we love, in the loss of our power through accident or

[1] Schiller, Frederich, "Ode to Joy," translated by Frederica Unger (New York: publisher unknown, 1959), p. 42.

[2] Tagore, Rabindranath, *Sadhana, The Realization of Life* (New York: Macmillan Publishing Company, 1916), pp. 99, 116.

illness, in the disappointment of our hopes. Just as day does not exist without night, nor life without death, joy cannot exist apart from sorrow. There is pain in life as well as pleasure, but we can accept the pain as long as we are not stuck in it. We can accept loss if we know we are not condemned to grieve continually. We can accept night because we know day will break, and we can accept sorrow when we know joy will again spring forth. But joy can spring forth only when our spirit is free. Unfortunately, too many people have been broken, and for them joy is not possible until they heal.

How did man lose his joyfulness? The Bible offers some understanding. It tells us that at one time man and woman lived in the Garden of Eden, which was paradise. Like all the other animals in this garden, they lived in a state of blissful ignorance. In the garden there were two trees, the fruit of which they were forbidden to eat: the tree of knowledge and the tree of life. The serpent tempted Eve to eat the fruit of the tree of knowledge, saying that it was good. Eve protested, saying that if she ate the forbidden fruit, she would die. But the serpent pointed out that she would not die since she would become like God, knowing good from evil. Eve then ate the fruit and convinced Adam to do the same. As soon as they did, they gained knowledge.

The story reveals how man became a self-conscious creature. The forbidden knowledge was a consciousness of sexuality. Every other animal is naked but none feels ashamed. All other animals are sexual but they are not self-conscious of their sexuality. This self-consciousness robs sexuality of its naturalness and spontaneity and, in effect, robs the human being of his innocence. The loss of innocence leads to guilt, which destroys joy.

The story is allegorical, but it describes the experience of every human being in the process of acculturation. Every child is born in a state of innocence and freedom and can thus experience joy. Joy is the natural state of a child, as it is of all young animals, as is obvious to anyone who has watched spring lambs gambol, jumping for joy.

Feeling the Life of the Body

Joy belongs in the realm of positive body feelings; it is not a mental attitude. One cannot make up one's mind to be joyful. The positive body feelings start from a baseline which can be described as "good." Its opposite is to feel "bad," which means that, instead of a positive excitement, there is the negative excitement of either fear, despair or guilt. If the fear or despair is too great, one will suppress all feeling, in which case the body becomes numb or lifeless. When feelings are suppressed, one loses the ability to feel, which is depression, a state that unfortunately can become a way of life. On the other hand, when the pleasurable excitement mounts from the baseline of good feeling, one knows joy. Should joy overflow, it becomes ecstasy.

When the life of the body is strong and vibrant, feeling, like the weather, is changeable. We can be angry one moment, then loving and crying the next. Sadness can change into pleasure just as the sun can follow rain. This change of mood, like a change in the weather, does not upset one's basic equilibrium. The changes take place at the surface and do not disturb the deep pulsations which provide the person with a sense of well-being. Suppressing feeling is a deadening process which diminishes the body's inner pulsation, its vitality, its state of excitement. For this reason, suppressing one feeling suppresses all others. If we suppress our fear we suppress our anger. Suppressing anger results in the suppression of love.

We human beings are taught early in life that certain feelings are "bad" while others are "good." It is actually so stated in the Ten Commandments. To love and honor one's father and mother is good, to hate them is bad. It is a sin to desire thy neighbor's wife, but if she is an attractive woman and we are vital men, such desire is perfectly natural.

It is important to note, however, that to have the feeling is not the sin; it is acting on the feeling that makes it a social issue. In the interest of social harmony we have to impose controls on behavior. "Thou shalt not kill" or "thou shalt not steal" are necessary restrictions when people live in large or small groups. Human beings are social creatures whose survival depends on the cooperative action of the group. Restrictions on behavior, which further the welfare of the group, are not necessarily injurious to the individual. Restrictions on feeling are another matter. Since they are the life of the body, judging feelings as good or bad is to judge the individual, not his actions.

To condemn any feeling is to condemn life. Parents often do this, telling a child that he or she is bad for having certain feelings. This is especially true of sexual feelings but also of many other feelings. Parents often shame a child for being frightened, which forces the child to deny his fear and act brave. But not feeling fear doesn't mean that one feels courage—only that one doesn't feel. No wild animal knows right from wrong, has a sense of shame or feels guilty. No animal judges its feelings or its actions—or itself. No animal living in nature has a superego or is self-conscious. It is free from internal constraints derived from fear.

Feeling is the perception of an inner movement. If there is no movement, there is no feeling. Thus, if one lets his arm hang motionless for several minutes, one loses the feeling of the arm. We say it "goes dead." This principle holds true for all feelings. Anger, for example, is a surge of energy in the body activating the muscles that would carry out the angry action. That surge is an impulse which, when perceived by the conscious mind, creates a feeling. Perception, however, is a surface phenomenon: an impulse leads to feeling only when the impulse reaches the surface of the body, which includes the voluntary muscle system.[3] There are many pulses in the body which do not result in feeling because they remain confined to the interior. We ordinarily do not feel the beating of the heart because the pulse does not reach the surface. Should the beating become very strong, its effect is felt on the surface of the body and we become conscious of our hearts.

When an impulse reaches a muscle, the muscle becomes set to act. If it is a voluntary muscle, the action is under the control of the ego and can be restrained or modified by the conscious mind. Blocking the action creates a state of tension in the muscle, which is energetically charged to act but is unable to do so by the restraining command from the mind. At this point the tension is conscious, which means that it can be released by withdrawing the impulse or by releasing it in a different form, such as slamming one's fist on the table rather than in someone's face. However, if the insult or injury which provoked the anger continues as a disturbing irritant, the angry impulse cannot be withdrawn. This is true of conflicts between parents and children since the latter cannot escape a parent's hostility. And in most cases the child has no means to discharge the impulse without provoking more anger and hostility from the parent. In this situation the tension becomes chronic and painful. Relief is possible only by numbing the area, rendering it immobile so that all feeling is lost.

Individuals who have suppressed their anger against their parents out of fear show marked tension in the muscles of the upper back. In many cases the upper back is rounded and up as it would be in a dog or cat who is prepared to attack. We could describe such a person as "having his back up" to indicate an

[3] Lowen, Alexander, *The Language of the Body* (New York: Macmillan Publishing Co., 1971).

angry attitude. But the individual is not in touch with his bodily attitude, nor with the potential anger that underlies it. It is frozen and he is numb. Such an individual may go into a rage over a minor provocation without sensing that he is venting a long-suppressed anger. Unfortunately such a rage does not release the tension because it is an explosive reaction and not a true expression of the underlying anger.

Such chronic muscular tensions are found throughout the body as signs of blocked impulses and lost feelings. The jaw is an area of chronic muscular tension which is so severe in some individuals that it constitutes a disease entity known as temporomandibular joint disease. The impulses which are blocked are crying and biting. One sets the jaw to maintain self-control in conditions where one could break down and cry or run in fear. When such control is conscious and can be surrendered at will it serves the person's well-being. Chronic tension in the jaw, on the other hand, cannot be released by a conscious effort, except momentarily, since it represents an habitual or characterological attitude of determination. Every chronic tension represents a limitation on the individual's ability to express himself. Most individuals in our culture suffer from considerable chronic tension in their musculature—in the neck, chest, lower back and legs, to name some areas—which binds them, restricting their grace in movement and destroying their ability to express themselves freely and fully.

Chronic muscular tension is the physical side of guilt because it represents the ego's injunction against certain feelings and actions. A few individuals who suffer from such chronic tensions actually feel their guilt, but most are not aware that they feel guilty nor of what their guilt is about. In a specific sense, guilt is the feeling of not having the right to be free, to do what one wants. In a general sense, it is the feeling of not being at ease in one's body, of not feeling good. When one doesn't feel good in one's self, the underlying thought is *I must have done something bad or wrong.* For example, when one tells a lie, one feels bad or guilty because one has betrayed his true self, his true feeling. One would naturally feel guilty about the lie. There are people, however, who don't feel guilty when they lie, but that is because they don't feel; they have suppressed feeling. On the other hand, one cannot feel guilty if one feels "good" or joyful. The two states —feeling good/joyful and feeling bad/guilty—are exclusive.

In most cases a forbidden fruit evokes mixed feelings. It tastes good, which is one reason it is forbidden. But because it is forbidden by the superego—that is, that part of the conscious mind which has incorporated parental dictates—we cannot surrender to the pleasure. This creates a bitter taste in our mouths which becomes the core of the feeling of guilt. Sexuality is, of course, the forbidden fruit in our culture, and almost all civilized individuals suffer from some measure of guilt or shame about their sexual feelings and fantasies. In narcissistic individuals there is a denial of and dissociation from feelings, with the consequence that they do not feel shame or guilt, but are also unable to feel love.[4] These individuals seem uninhibited and free in their sexual behavior but their freedom is external, not internal—in action, not in feeling. Their sexual actions constitute a performance, not a surrender to love. For them sex is an act, not a joyful experience. Without the inner freedom to feel deeply and express one's feelings fully there can be no joy.

Inner freedom is manifested in the gracefulness of the body, in its softness and aliveness. It corresponds to a freedom from guilt, from shame and from self-consciousness. It is a quality of being that all wild animals possess but that is absent in most civilized beings. It is the physical expression of innocence, of a

---

[4] Lowen, Alexander, *Narcissism, The Denial of the True Self* (New York: Macmillan Publishing Co., 1985).

way of acting that is spontaneous, without guile and true to the self.

Unfortunately, a lost innocence cannot be recovered. Having gained knowledge of right and wrong and of sexuality, are we doomed to be sinners? Do we have to live a life of guile, manipulation and self-deceit? We should remember that all religions preach a salvation. We are not doomed to hell or even purgatory, although many people seem to exist on those levels in this life. Salvation always involves a surrender to God, an abandonment of one's egotism, the commitment to a moral life. But this is easier said than done. We have lost touch with God because we have lost touch with the God within us—the whirling spirit that animates our being, the pulsing center of our inner self that lights up our being and gives meaning to our life.

In this book I will describe the torment and suffering that my patients have described and which brought them to therapy. To connect with our inner God is the goal of therapy. That God resides in the natural self, the body which was created in the image of God. The natural self lies buried deep within the body under layers of tensions representing superego injunctions and suppressed feelings. To reach that self a patient has to take a journey backward in time to one's earliest years. It is a painful journey because it awakens frightful memories and evokes painful feelings. But as the repression is lifted and the suppression of feeling eased, the body that God created slowly comes fully alive.

The voyage of self-discovery that constitutes the therapeutic process cannot be taken alone. Like Dante in *The Divine Comedy*, the solo voyager is lost and confused. Dante, in his distress at finding himself lost in a forest and feeling threatened by wild animals, called upon Beatrice, his protectress in heaven, for aid. She sent him Virgil, the Roman poet, to act as his guide, for the way home led through Hell, which posed dangers to the traveler. Virgil could help Dante pass safely through this frightening area since he had traversed it himself earlier. With Virgil's help Dante makes a safe passage through Hell, after which he also goes through Purgatory and then enters Paradise. In the therapeutic process the guide is a person who has made a similar voyage of self-discovery through his own hell. To be an effective guide in an analytic therapy, the therapist must have undergone a full analysis himself, one that ended in his own self-realization.

For the patient in therapy, hell is the repressed unconscious, the underworld in which are buried the terrors of the past—despair, torment, mania. If the patient descends into this dark world, he will experience the painfulness of his buried past; he will relive the conflicts that he could not handle and he will discover a strength that he had dreamed of but not believed possible. Initially the strength comes from the guidance, support and encouragement of the therapist, but it becomes the patient's strength as he finds that his terrors are childhood fears which an adult can deal with. Hell exists only in the darkness of night and death. In the light of day—that is, with full consciousness—one does not see any real monsters. Wicked stepmothers turn out to be angry mothers who terrified the child. Feelings which are thought to be shameful, dangerous and unacceptable turn out to be natural reactions to abnormal situations. Slowly the patient repossesses his body and, with it, his soul and his self.

I have pointed out elsewhere that the unconscious is that part of the body one doesn't feel.[5] There are large parts of our body that we cannot feel. We have no consciousness of the functioning of our blood vessels, nerves, endocrine glands, kidneys, etc. Some Indian fakirs seem to be able to deepen their awareness so that they can sense those organs, but that is not the way consciousness typically works. Consciousness is like the tip of the iceberg that projects above the surface of the sea, but it also includes the part

[5] Lowen, A., *The Language of the Body*.

that is just below the surface, which also can be seen. In persons
with emotional problems or conflicts there are areas of the body
within the normal range of consciousness which are not sensed
because they have been immobilized by chronic tension. Immobilization blocks threatening impulses but also deadens that part
of the body resulting in a loss of that part of the self. These areas,
therefore, represent emotional conflicts which have been repressed into the unconscious. For example, most individuals do
not sense the tension in their jaws and are not aware that this
tension represents the suppression of biting or crying impulses.
These conflicts represent the repressed unconscious. They constitute the underworld in which are buried those feelings which
the ego or conscious mind believes are dangerous, shameful and
unacceptable.

Like the souls in Hell, these feelings which are dead to the
conscious mind live on in an underworld of torment. Occasionally the torment rises to consciousness, but since it threatens survival it is pushed down again. We can survive if we live on the
surface where we can control feeling and behavior, but this entails a sacrifice of deep feelings. Living on the surface in terms
of ego values is a narcissistic way of life which proves to be
empty, and generally results in depression. Living in the depth
of one's being can be painful and frightening at first, but it also
can be fulfilling and joyful if we have the courage to go through
the hell within to reach paradise.

The deep feelings we have buried are those which belong to
the child we were, the child who was innocent and free and who
knew joy until his spirit was broken by being made to feel guilty
and ashamed of his natural impulses. That child still lives in our
hearts and in our guts but we have lost contact with it, which
means we have lost contact with the deepest part of ourselves.
To find ourselves, to find the buried child, we must go down
into these deep areas of our being, into the darkness of the un-

conscious. We must brave the fears and dangers of this descent, and for that we need the help of a therapist-guide who has completed this journey in his own process of self-discovery.

These ideas parallel mythological thinking, in which the diaphragm is equated with the surface of the earth. The half of the body above the diaphragm lies in the light of day; the part below, namely the belly, lies in the darkness of night and the unconscious. The conscious mind has some control over the processes of the upper half of the body, but we have little or no control over processes in the lower half, which includes the functions of sexuality, excretion and reproduction. This part of the body is closely connected with man's animal nature whereas the functions of the upper half are more subject to cultural influences. The simplest way to describe the difference is to say that we eat as human beings but defecate like animals. Perhaps because the lower half of the body is more associated with our animal nature, its functions, especially sexuality and movement, are capable of yielding experiences which are highly enjoyable, even ecstatic.

CHAPTER 2

THE SURRENDER
TO THE BODY

The Surrender of the Narcissistic Ego

The idea of surrender is unpopular with the modern individual who sees life as a struggle, a fight, or at least a competitive situation. For many people, life aims at some achievement, some success. One's identity is often tied to one's activity rather than one's being. This is typical of a narcissistic culture in which the image is more important than the reality. In fact, for many people it replaces reality.[1] In a narcissistic culture success seems to confer self-esteem, but only because it inflates one's ego. Failure has the opposite effect because it deflates the ego. In this atmosphere the word "surrender" is equated with being defeated, but it is really only a defeat of the narcissistic ego.

Without a surrender of the narcissistic ego one can't surrender to love. Without such a surrender joy is impossible. Surrender doesn't mean the abandonment or sacrifice of the ego. It means that the ego recognizes its role as subservient to the self—as the

---

[1] See A. Lowen, *Narcissism, The Denial of the True Self*, for an in-depth analysis of the narcissistic personality.

organ of consciousness, not the master of the body. We must recognize that the body has a wisdom stemming from several billion years of evolutionary history which the conscious mind can imagine but never grasp. The mystery of love, for example, is beyond the reach of scientific knowledge. Science can make no connection between the heart as a pump to send blood through the body and the heart as the organ of love, which is a feeling. Wise men have understood this seeming paradox. Pascal's statement that "the heart has its reasons which reason will never know" is true.

It is not true that mind and body are equal, as some persons claim. Their seeming equality is the result of the limited vision of the conscious mind, which sees only the surface of things. Like our view of the proverbial iceberg, we see only a little more than ten percent of its mass. The part concealed in darkness, the unconscious part of our body, is what keeps our life flowing. We do not live by our will. The will is impotent to regulate or coordinate the complex biochemical and biophysical processes of the body. It is unable to affect the body's metabolism upon which our life depends. And that is a very reassuring concept, for if the reverse were the case, life would collapse at the first failure of the will.

Consider the development of the embryo into a human being, a process that awes the human mind. That tiny organism, the fertilized egg, "knows" what it has to do to potentiate its inherent possibility to become a human being. It is awesome. And yet we human beings have the arrogance to think that we may know more than nature. I place my faith in the power of the living body to heal itself. This is not to say that we cannot help the healing process. But we cannot substitute for it. Therapy is a process of natural healing in which the therapist supports the body's own healing function. It is not the doctor who tells the body how to repair a broken bone, and it is not the doctor who orders the skin to regenerate itself after a tear or a cut. In many cases the healing will take place even without the support of a medical person.

I have asked myself why this doesn't happen with emotional or mental illness. If we get depressed, why don't we heal spontaneously? In fact, some people do get over a depressive reaction spontaneously. Unfortunately, in most cases the depression tends to recur because the underlying cause persists.[2] That cause is the inhibition of expressing one's feelings of fear, of sadness and of anger. The suppression of these feelings and the concomitant tension reduce the motility of the body, resulting in a state of reduced or depressed aliveness. Coupled with this is the illusion that one will be loved for being good, subservient, successful, and so forth. This illusion serves to maintain the spirits of the individual during the struggle to win love, but, since true love can't be earned or won by any performance, the illusion collapses sooner or later and the individual becomes depressed. The depression will lift if the individual can feel and express feeling. Getting a depressed patient to cry or become angry will lift him out of the depression—at least temporarily. Expressing feeling releases tension, allowing the body to recover its motility, thereby increasing its aliveness. This is the physical side of the therapeutic process. On the psychological side, one needs to uncover the illusion and to understand its origin in childhood and its role as a survival mechanism.

All patients suffer from some illusion to varying degrees. Some have the illusion that wealth brings happiness, or that fame insures love or that being submissive protects one from possible violence. We develop these illusions early in life as a means of

[2] For an analysis of the causes of depression see Lowen, Alexander, *Depression and the Body* (New York: Coward, McCann & Geoghegan, Inc., 1972; Arkana, 1993).

surviving a painful childhood situation and we are afraid as adults to surrender them. Perhaps the biggest illusion of all is the belief that the conscious mind controls the body and that if we change our thinking, we can change our feelings. I have never seen it work, although the illusion that the mind is all-powerful can buoy up one's spirits temporarily. But this illusion, like all others, will collapse as the person runs out of energy, and the result will be depression.

Illusions are ego defenses against reality, and while they may spare one the pain of a frightening reality, they make us prisoners of unreality. Emotional health is the ability to accept reality and not run away from it. Our basic reality is our body. Our self is not an image in our brain but a real, living and pulsating organism. To know ourselves we have to feel our body. The loss of feeling in any part of the body is the loss of part of the self. Self-awareness, the first step in the therapeutic process of self-discovery, is the feeling of the body—the whole body, from head to toes. Many individuals under stress lose the feeling of the body. They dissociate from the body to escape reality, which is a schizophrenic type reaction and constitutes a serious emotional disturbance. But almost all people in our culture dissociate from parts of their body. Some have no feeling in their back. This is especially true of individuals who can be described as having no backbone. Others lack feeling in their guts. These individuals will manifest a lack of courage. Every part of the body contributes to our sense of self if we are in touch with it. And we can only be in touch with it if it is alive and mobile. When every part of the body is charged and vibrant, we feel vibrantly alive and joyful. But for that to occur we need to surrender to the body and its feelings.

Surrender means letting the body become fully alive and free. It means allowing the involuntary processes of the body, like respiration, full freedom of action and not controlling them. The body is not a machine that one has to start or stop. It has a "mind" and knows what to do. In effect what we are surrendering is the illusion of the power of the mind.

The best place to begin is with breathing. This is the basis of the technique that Reich employed in his therapy with me. Breathing is perhaps the most important bodily function, since life depends so much upon it. It has the distinction of being a natural, involuntary activity but at the same time one that is subject to conscious control. In ordinary circumstances one is not conscious of breathing. However, when one has difficulty getting enough air, as in high altitudes, one becomes conscious of laboring to breathe. For people with emphysema breathing is a constant painful struggle to get enough air.

Emotional states directly affect one's breathing. When a person is very angry, his breathing becomes more rapid to help him mobilize more energy for aggressive action. Fear has the opposite effect, causing a person to hold his breath because action is suspended in a state of fear. If the fear becomes panic, as when a person desperately tries to escape a threatening situation, breathing becomes rapid and shallow. In terror, one hardly breathes at all because terror has a paralyzing effect upon the body. In a state of pleasure, breathing is slow and deep. However, should the pleasurable excitement mount to joy and ecstasy, as in the sexual orgasm, breathing becomes very rapid but also very deep in response to the heightened pleasurable excitement of the sexual discharge. Studying a person's breathing allows the therapist to understand his emotional state.

Though I described my therapy with Reich in an earlier book, I'll recount some of my experiences again to illustrate the concept of surrender. I lay on a bed wearing just a pair of shorts so Reich could observe my breathing. He was sitting facing the bed. His simple instruction was to breathe, which I proceeded to do as I

normally would, while he studied my body. After ten or fifteen
minutes he remarked, "Lowen, you're not breathing." I replied
that I was. "But," he said, "your chest isn't moving." It wasn't.
He asked me to place my hand on his chest to sense its motion.
I felt the rise and fall of his chest and decided to mobilize my
chest with each breath. I did this for some time, breathing
through my mouth, feeling quite relaxed. Reich then asked me
to open my eyes wide, and as I did so I uttered a loud, sustained
scream. I heard myself screaming but I had no feeling attached
to it. It was coming from me but I was not connected to it. Reich
asked me to stop the scream because the windows of the room
were open to the street. I resumed my breathing as before as if
it had not happened. I was surprised by the scream but not emotionally affected. Then Reich asked me to repeat the action of
opening my eyes wide and again I screamed without any emotional connection to it.

We met three times a week, but nothing dramatic happened
in the next two to three months. Reich encouraged me to let go
and breathe more freely, which I tried to do. Despite my efforts,
Reich told me that my breathing was not free, that I was consciously doing it as an exercise and not just letting it happen.
Unconsciously I was controlling my breathing so that nothing
more would happen, but I didn't know this then. I tried to let
go of my control, to give in to my body and its involuntary
processes, but this was difficult for me to do. Breathing more
fully, though consciously done, led to symptoms of hyperventilation. Strong tingling sensations, known as paraesthesias, developed in my hands and arms. At one point my hands froze in a
Parkinsonian contracture. They were ice-cold, like claws, and
paralyzed. But I was not frightened. I breathed more quietly,
and slowly the contracture released and the paraesthesias disappeared. My hands became warm again. After several sessions in

which the deeper breathing produced this hyperventilation syndrome, the reaction disappeared. My body had adapted to the deeper breathing and was becoming more relaxed.

Shortly thereafter the therapy was interrupted for Reich's summer vacation. When we resumed in the fall, it was back to giving in and breathing spontaneously. In the course of this next year of therapy several important events occurred. In one of them I relived an infantile experience which explained the screams of my first session. As I lay on the bed breathing, I had an impression that I would see an image on the ceiling. Over several sessions the impression became stronger. Then the image appeared. I saw my mother's face. She was looking down at me with very angry eyes. I felt that I was a baby about nine months old lying in a carriage outside the door of my house and crying for my mother. She must have been involved in some important activity, for when she came out, she looked at me with such anger that I froze in terror. The screams I couldn't utter then burst forth in my first therapy session, thirty-two years later.

On another occasion I had the unusual experience of feeling myself moved by some inner force. My body began to rock and, from my lying-down position, I sat up, then stood up. Facing the bed, I began to hit it with both fists. As I did so I saw the face of my father and I knew that I was hitting him because he had spanked me when I was about seven or eight. When I asked him about this incident later, he confirmed it, explaining that I had stayed out late, worrying my mother, and that she had demanded the punishment. The amazing thing about this experience was that my movements were not consciously made. I did not decide to get up and hit the bed. My body acted on its own just as it did when I screamed.

During the second year of my therapy with Reich my breathing was much freer. Although I could not surrender fully to my body, its motility increased considerably. As I lay on the bed breathing, vibrations developed in my legs as I gently moved them apart and together. These vibrations indicated that an energetic current was flowing through them which felt very pleasurable. I was also able to experience these vibrations in my hips as they became more alive. These vibrations stemmed in part from the release of tension in the muscles of these areas, but in part it is a natural phenomenon of life. Living bodies are vibratory systems, dead bodies do not move. Despite the two breakthroughs and the increasing aliveness of my body, I was not able to surrender fully to the point where the orgasm reflex would occur. At this point Reich suggested that we terminate the therapy since it seemed to have reached a dead end.

This suggestion had a powerful effect on me. I broke down and sobbed deeply. Stopping the therapy represented failure and the defeat of my dream to achieve sexual health. I expressed this feeling to Reich and also how much I wanted his help. Asking for help was also difficult for me. I believed I had to do it alone and by myself. But surrendering to the body and its feelings was something I could not do. Doing is the opposite of surrendering. Doing is an ego function whereas surrendering to the body requires an abandonment of the ego. I would not have regarded myself as an egotistic or narcissistic individual, but I have since learned that was an important aspect of my personality. I would not or could not break down and cry (unless pushed to the extreme; that is, threatened by the loss of my heart's desire), for, unconsciously, I was determined to succeed.

Recognizing the significance of my breakdown, Reich agreed to continue the therapy. Following this episode I was able to give in more fully and my breathing became freer and deeper. When it again came to Reich's summer vacation, he suggested that I take an entire year's absence from therapy and return the next fall. I welcomed the suggestion since I wanted a break from the effort to get well. The breakdown that the crying represented allowed me to surrender more fully to my feeling of love than I had been able to do previously. I had fallen in love with a young woman about a year earlier but the relationship was not solid. At one point, when it seemed that it would end, I broke down again and cried very deeply, expressing my love for her. Following this episode I had the most intense and pleasurable sexual experience I have known, which I recognized stemmed from this surrender to my deepest feeling. During the following year I was married to this lady, and, I might add, still am.

When I resumed my therapy after the year's interruption, my ability to give in to my body's involuntary actions improved greatly and it was not long before the orgasm reflex developed. I felt excited and joyful. I felt transformed, but it didn't hold up. Transforming experiences reveal the possibility of joy and are, therefore, meaningful and precious, but rarely do they go deep enough to have a lasting effect. For that, one has to work through the conflicts stemming from the past that are deeply structured in the personality both psychologically and physically. Too many of my problems had been left unresolved in my therapy with Reich to allow me to be free and fully open to my feelings. Nevertheless, the experiences I had in my therapy convinced me that the way to joy could be reached only by surrendering to the body.

After several years of study which led to receiving my medical license, I returned to my practice, using the technique which I had learned from Reich. The patient would lie on a bed, relaxed and breathing, while I encouraged him to give in to his breathing and surrender to his body. We also talked about his life and his problems. But nothing much happened. Sitting in a chair watching him, I felt a need to stretch over the back of my chair to get a deeper breath. It occurred to me that this is what my patients needed to do. In the kitchen of the office there was a three-step kitchen ladder stool. I rolled up a blanket and tied it to the stool.

FIGURE 1

Then I had the patient lie on his back over the stool, his arms reaching back to a chair, as shown in Figure 1. The effect was very positive. The patient's breathing deepened appreciably because of the stretch. I could observe the respiratory wave and note where it was blocked.

Since then the use of the bioenergetic stool has become a regular part of my therapeutic approach. In the forty years since it was first introduced into bioenergetic analysis, I have learned how to increase its effectiveness by having the patient use his voice while on the stool. I will describe how I coordinate the voice with breathing in the next chapter.

Another important change I made in the Reichian technique was the use of specific body exercises designed to help a patient gain better awareness of his body, fuller self-expression and more self-possession. Prior to meeting Reich I had been an athletic director. My considerable experience with exercises showed me that they could have a strong effect upon one's feelings and state of mind. I developed the therapy-related exercises originally to increase the motility of my own body, then began to design new ones to deal with the specific emotional problems I observed in the body of a patient. Many of these exercises involve the expression of feeling. They will be described in succeeding chapters.

The first exercise I did to increase the feeling in my legs, and thus increase my sense of security, is called the bow. It is actually a well-known position since it is also part of the Chinese exercise program called Tai Chi Chuan, but I did not know this in 1953 when I first used it. I stood with feet widely spread, knees bent and body arched slightly. To maintain the arch I placed my fists into the small of my back. This position gave me a secure feeling of being more in touch with the lower part of my body. The position also facilitated deeper breathing, which is one reason the Chinese use it. Sensing my way, I reversed the position, bending forward with my fingers touching the ground, my feet about twelve inches apart and turned slightly inward. In this position I felt close to the ground and to my legs and feet. Then, if I kept the weight of my body over my feet and slowly straightened my knees without locking them, my legs would generally begin to vibrate. Figure 2 illustrates this position.[3]

In the course of my therapy with Reich I had experienced vibrations in my body, particularly in my legs and hips, as I was lying on the bed breathing. They were an involuntary action which developed in response to the wave of excitation that flowed through my body. Individuals who are unable to let go because their bodies are too tight find it very difficult to allow the vibrations to occur. However, doing these exercises regularly helps the person feel the pleasure of letting the body become more alive. I found that vibrations were also induced by gentle movements

[3] These exercises and others are fully described in the book, *The Way to Vibrant Health* by A. and R. L. Lowen (New York: Harper & Row, 1977; International Institute for Bioenergetic Analysis, 1992).

FIGURE 2

of the legs and always resulted in pleasurable sensations in these areas. But in Reich's therapy these movements were not considered deliberate exercises which one could use regularly as part of the therapeutic program. Today the above exercises and others are a regular part of the bioenergetic program to help an individual feel more grounded, more connected to his body and to reality. They did that for me and I continue to do them regularly for myself as well as using them with my patients.

Grounding and Reality

The surrender to the body is associated with the giving up of illusions and coming down to the ground and to reality. The individual who is strongly connected to reality is said to "have his feet on the ground," meaning he feels the connection between his feet and the ground he stands on. Individuals who are hung up or uptight do not feel this contact with the ground because their feet are relatively numb—they may know their feet touch but they have no sensation of the contact. They have withdrawn this energy-as-excitation from the lower part of the body as a reaction to fear. Where the fear is very great the person may actually withdraw all feeling from his body, limiting his consciousness to his head. He will then live in a fantasy world, which is common in autistic or schizoid children and adults. Many persons live in their heads more than in their bodies to avoid sensing the frightening and painful feelings in the body. Some actually split off and dissociate from the body in situations of extreme fear. Their consciousness moves out of the body and they experience themselves as looking at the body from above. This is a schizophrenic-type reaction and represents a break with reality. One of my patients reported feeling himself up at the ceiling looking down at his body as it lay on the bed. He was, of course, a very disturbed individual.

Contact with reality is not an all-or-nothing condition. Some of us are more in contact with reality than others, who are more split off. Since contact with reality is the condition of sanity, it is also the condition for emotional and physical health. Many people, however, are confused about what reality is since they equate reality with the cultural norm rather than with what they feel within their body. Of course, when feeling is absent or reduced, one looks beyond the self for the meaning of life. Individuals whose bodies are alive and vibrant can feel the reality of their being, and can be said to be a feeling person. How alive one is and how much feeling one has is a measure of one's contact with reality. Feeling individuals are "down-to-earth" people. We describe such individuals as being "grounded."

To be grounded means to feel one's feet on the ground. To feel the ground, one's legs and feet have to be energetically charged. They must be alive and mobile, that is, showing some spontaneous and involuntary movement such as vibration. The vibration doesn't have to be intense; it can be quiet—just a hum like the purr of a high-powered car. But when there is no hum in a car, we know that the motor is dead. When a person's feet look lifeless and when his legs look still and immobile, we know that he has no feeling-contact with the ground. When a person's legs and feet are fully alive, he can sense a current of excitation flowing through them, exciting them, warming them and vibrating them. I was consulted by a schizophrenic young woman who had walked to my office through snow-covered streets wearing only light sneakers. Her feet were cold and blue but she had no feeling of pain and no awareness of their condition. They were numb and almost lifeless. She was, obviously, ungrounded and completely out of touch with her body.

Grounding is an energetic process in which there is a flow of excitation through the body from head to feet. When that flow is strong and full, the person feels his body, his sexuality and the ground on which he stands. He is in contact with reality. This flow of excitation is associated with the respiratory waves, so that when breathing is free and deep, the excitation flows similarly. If the breathing or the flow is blocked, the person does not feel his body below the block. If the flow is restricted, feeling is reduced. Because the flow of excitation pulses—flowing downward into the feet, then upward into the head, like the swings of a pendulum—it excites the segments of the body, head, heart, genitals and legs. Since the wave of excitation traverses the pelvic area as it flows downward, any major sexual disturbance will block the flow to the legs and feet. When an individual is ungrounded, his sexual behavior is likewise ungrounded, that is, dissociated from feeling in the rest of the body.

Since to be grounded means to stand on one's own feet, it also denotes the state of independence and maturity. By the same token the standing position represents a more adult position than one lying on a bed, which has a more infantile quality. Thus, it is easier for a patient to regress to an infantile position when lying down than when standing. This explains why experiences such as the orgasm reflex, which a patient could have during a therapy session lying on a bed, does not necessarily translate into changes in adult behavior. The orgasm reflex is a valid but not necessarily an absolute criterion of health. The individual must also be fully grounded. We must recognize that the feelings of a child, while similar to those of an adult, are not identical. A child's anger is not the same as that of an adult, nor is its sadness. Adult love differs from that of a child, not in its essential quality, since that is a function of the heart, but in its breadth and extension, which are determined by the total body. This does not mean that babies and young children are not grounded. They are grounded through their connection with their mother as a representation of the earth, but they are not directly connected to the ground until they become fully able to stand on their own feet.

This analysis helps one understand the appeal of a cult which demands of its members the surrender of their egos to the cult leader. The surrender to a leader amounts to a regression to childhood, and involves an abdication of power and responsibility. Protected by the leader and unhindered by the need to choose between right and wrong, the cult member has a feeling of freedom and innocence. As a result, he experiences a sense of joyfulness which strengthens his commitment to the cult. The question arises as to whether his feeling of joy is an illusion or reality. Illusions can produce real feelings but they do not hold up when the illusion collapses, as all illusions inevitably do. In the case of the cult, the illusion is that the leader is the all-loving, all-powerful father who will take care of the cult members as a good father would take care of his children. The reality is the contrary since cult leaders are narcissistic individuals who need a following to support their grandiose self-images. They also need power over others to compensate for their impotence. Of course cult leaders only attract those who are unconsciously looking for a powerful father/leader.

Some elements of the relationship between the cult leader and his followers were present in my relationship with Reich, although I never became a follower. At the time I broke down and cried at the prospect of my therapy with him ending in failure, I was aware of how much I wanted his protection and looked upon him as the good and powerful father. The threatened failure of the therapy represented the loss of that hope. My crying was partly for the loss of that hope, but it was also an expression of my sadness at not having had the kind of father who could have provided the support I needed to feel free and joyful. My defense against the pain and sadness of this lack was to adopt the attitude that I didn't need help and that I could do it myself. This is the way I operated in the world and to all appearances it seemed true. But on a deeper level, it didn't work.

A cult did develop around Reich in the years after my therapy with him ended. I never became part of the group which surrounded Reich from 1947 to 1956 and who looked upon him as all-knowing and all-powerful. In part this was because I left for Europe in 1947 to study medicine at the University of Geneva, which took me out of his circle. More important was my wife's influence. She had a very strong distrust of any closeness based on submission or on uncritical acceptance of another human being as superior, all-knowing or all-good. She saw too many people close to Reich at that time who had surrendered their independence and mature judgment to gain some intimacy with the great man. I could see it, too. Having said this I would add that, in my view, then and now, Reich was a great man in many respects. His understanding of the emotional problems of human beings, his perception of the underlying unity in all of nature and the clarity of his thinking set him above all others in his field. But he was not all-knowing and he had many personal problems which handicapped his work and his life.[4]

The therapeutic situation necessarily fosters an attachment to the therapist, who can be legitimately regarded as a substitute father- or mother-figure. One goes to a therapist because one needs help in the form of acceptance, understanding and support. If the therapist takes a personal interest in the patient, the latter can become easily attached, dependent and in love with the therapist. This attachment to the therapist, positive as it is in many respects, weakens the patient's awareness of his need for independence and leads to his being "hung up" on his therapist and in an ungrounded state. It is also recognized that the patient will transfer to the therapist all the feelings he had about his or her own parent, both positive and negative. Positive feelings encourage submission and allow a patient to regress to a more infantile or childlike position, which facilitates the expression of feelings that had been denied and suppressed in childhood—namely, feelings of love. The expression of these feelings can lead to a sense of freedom and feelings of joy but, unless the negative feelings such as distrust and anger are also expressed, the good feelings do not hold up. They become undermined by the underlying negativity and despair which has not been resolved. These negative feelings, if not fully worked through in the therapy, undermine the initial surrender and leave the patient bitter and frustrated. The same thing happens in love relationships, where the joy of the initial surrender to the loved partner is undermined by unresolved hostilities stemming from childhood. As we shall see in the next chapters, these negative feelings include a deep

---

[4] See Myron Sharaf's excellent biography, *Fury on Earth* (New York: Da Capo Press, 1994), which documents Reich's achievement but also portrays his personal conflicts and problems.

despair and a murderous rage which must be experienced and lived-through in the therapy situation if the patient is to become free. The patient's fear of these feelings constitutes the backbone of this resistance to the surrender to the body, the self and life.

Every analytic therapist is aware of the need to bring these negative feelings to consciousness so that they can be worked through. Reich had made it a practice, when I was his patient, of asking me at each session if I had any negative thoughts or feelings about him. I recall denying that I did, which was the truth as far as my conscious awareness was concerned. Having become a "follower," I gave up my critical attitude, making possible my surrender to him and, through him, to my body. It was only after I separated from the Reichian movement because it had failed to give me what I needed that I became critical of Reich. What it had failed to give me was an in-depth understanding of my character. Reich's failure could be attributed to the fact that this therapeutic work with the body was not as deep and thorough as it should have been. One should remember that my therapy with Reich took place fifty years ago, at a time when an understanding of the energy dynamics of the body and personality were not as developed as they are today in Bioenergetic Analysis.

This development stemmed from a change in the position of the patient during the therapy from a supine or sitting one to a standing one. In classical psychoanalysis the patient lies on a couch and the focus is on the words he utters. Thoughts are the main material of the analytic process, while the quietness and passivity of the analytic situation eliminates or diminishes all other forms of self-expression. In my work with Reich I was also in a lying position, which, because it was passive, allowed me to regress to infantile or childhood states, thus facilitating the recovery of early memories. But words were not the main avenue of expression. Reich's attention was focused on how I breathed and what was happening on the body level. I was seen as well as heard, which greatly enlarged the therapeutic screen. Lying on the bed, I had my knees bent so I could sense my feet in contact with the bed, but the position was one of helplessness. On the other hand, when a patient stands, he takes an adult position that allows the focus to shift to the present, where his problems now are. The therapist can see from the patient's stance how he holds and presents himself to the world.

The most common stance that I have seen is an expression of passivity. The individual stands with his knees locked and his weight on the heels of his feet as if he is waiting to be told what to do. He is so unbalanced in this position that a slight push will topple him backwards. One senses from this stance that the person was trained to be good and obedient as a child. Having the person bend his knees slightly and move his weight forward to the balls of his feet changes the expression of his stance so that he now looks more aggressive, that is, prepared to move forward or into action. The standing position allows the therapist to evaluate how well- or poorly grounded a patient is, physically in relationship to the floor and psychologically in relation to his body.

In bioenergetic therapy a patient does not always stand. At the beginning of a session, patient and therapist sit facing each other so that the former can talk about what is happening in his life. From there the patient can use either the standing or lying-down position to work with his feelings. Sadness, for example, is generally more easily expressed when one is lying down, while the expression of anger is more difficult in this position. Hitting the bed to feel and express anger is used by many therapists, often without a full understanding of body language. I am referring to the practice of doing the hitting from a kneeling position. The position of kneeling denotes a submissive attitude, which contradicts the intention of the hitting action. One can get angry in a sitting position, but in that case the expression of anger is limited to words and gestures. Watching a person hit the bed from a standing position, one can observe how well the action is grounded in the reality of the feeling of anger. The patient whose hitting is unfocused and rageful rather than focused and angry has no feeling in his legs and feet to keep him connected to his body and the ground. The expression of rage does little to discharge the tension that keeps the patient hung up and out of touch with his reality.

Early in my practice I worked with a psychologist who had been severely depressed. He made such an excellent recovery that his wife consulted me about her problems. She said, "You were the only therapist who was able to get my husband on his feet again." I answered that I did it by putting him on his feet. This doesn't mean that putting a person on his feet literally will overcome the depression, but it is a move in that direction. Keeping a person just talking while sitting in a chair or lying on a couch handicaps the therapeutic process, in my opinion.

If the feeling of joy is to be an attribute of one's life, it cannot be dependent on some special experience. I am sure we have all known some moments of joy as a result of the breakthrough of a strong emotion, resulting in a feeling of liberation or freedom. It is like the sun breaking through clouds for a short time and then being covered again. Admittedly, it can't be sunny all the time, but we would like it to be sunny most of the time. Too many people live in the dark shadows of their past caused by frightening images which are not clearly seen. These images haunt the unconscious mind, producing disturbing dreams at night and vague anxieties during the day. Psychoanalysis was developed as a technique to bring these repressed memories to consciousness, so that the feelings associated with them could be expressed and discharged. I believe this is essential to every therapy. Before the sun comes up to cheer and warm us it is preceded by the light of daybreak. In analysis this is called insight, which one gains when the light of consciousness dispels the darkness in the soul.

As an analytic therapy, Bioenergetic Analysis recognizes the importance of the doctrine Know Thyself. In this work the self is seen not only as a reflection in the mind but also as the bodily self. Since the bodily self is more evident and objective than the reflection in a person's mind, which is subjective, getting to know one's self is a matter of getting in touch with the body. Many people are not in touch with their bodies, or at most feel only limited parts. They are not grounded in the reality of their bodies. The parts that one isn't in touch with contain the frightening feelings which are the counterpart of the frightening images in the mind. For example, most people do not feel their backs despite the fact that the back plays an important role in "backing up" the individual and supporting him when he is under pressure. This function is related to "having a backbone," that is, not to be a worm or a wimp. The backbone can serve this function only when it is experienced by the individual as an alive, energetic structure. If it is too weak or too pliable, the individual will lack the ability to "back up" his position and he will be seen by others as weak. If his backbone is too rigid, he could find himself immobilized in a posture of resistance which blocks his capacity to respond to life or love. I met a man some years ago who suffered from a condition known as ankylosing spondylitis—a rheumatic disease in which the backbone becomes frozen, almost as if it were a solid bone. He could not turn his head more than a few degrees to either side. It was painful to see, but I am not sure if he felt the pain. If he did, he never cried about it. His history included a very powerful, domineering father of whom he was literally scared stiff. But how did the backbone get involved in his struggle? If he had folded before his father's aggression, he would have been a "wimp" (no backbone). As a boy, he could not resist his father openly. He could only resist him internally by stiffening his backbone. This unconscious action preserved his internal integrity at the cost of his mobility and joy. It was sad but he wasn't sad; he had become frozen, and could not feel his body.

When a person lies over the bioenergetic stool breathing, he can sense the quality of his back. He can feel its state of tension or its weakness. Chronic tension is the physical equivalent of fear. Since fear immobilizes an individual, immobilization by chronic tension equals fear. Sensing the rigidity or the tension can help him become aware of his fear, which releases the repressed memories of his childhood. Lying over the stool, many patients expressed the fear that their back could break and then recalled that as a child they were afraid that their father would break their back if they defied him. This awareness enabled them to feel their anger, which was also blocked by the tension in the muscles of the back. Expressing the anger by then hitting the bed, for example, released the tension, restoring to the back both its flexibility and strength.

To whatever degree a person is out of touch with a part of his body, he is out of touch with the feeling related to the mobility of that part. A tight jaw and a tight throat will cut off feelings of sadness because one can't cry. If the overall body is rigid, the individual will have no feelings of tenderness. On a deeper level, many people lack feelings of love because their hearts are locked in a rigid thoracic cage which blocks both the awareness of the heart and the expression of heartfelt feelings.

The goal of therapy is self-discovery, which implies the recovery of one's soul and the liberation of one's spirit. Three steps lead to that goal. The first is self-awareness, and that means to sense every part of one's body and the feelings that can arise in it. It is surprising to me how many people are not aware of the expression on their face and in their eyes although they look at themselves every day. Of course, the reason they don't see their expression is that they don't want to see it. They believe they can't face it and that others can't either. So they wear a mask, a fixed smile which proclaims to the world that everything is all right when it isn't. When they drop the mask one generally sees an expression of sadness, pain, depression or fear. As long as they wear the mask, they cannot feel their face since it is frozen in a fixed smile. Feeling this sadness, pain or fear is not joyful, but if these suppressed emotions are not felt, they cannot be released. One is imprisoned behind a facade which cuts off the sun from reaching one's heart. When a person steps out of his dim cell, the sun may be too blinding, but when one gets used to it, one would not want to live in that dark place again.

The second step to self-discovery is self-expression. If feelings are not expressed they become suppressed, and one loses contact with the self. When children are forbidden to express certain feelings, like anger, or are punished for expressing them, these feelings are hidden and eventually become part of the shadowy underworld of the personality. A lot of people are terrified of their feelings, which they regard as dangerous, frightening or crazy. Many individuals have an unconscious murderous rage which they feel they must keep buried out of fear of its destructive potential. In a few individuals that rage is conscious. Such rage is like an unexploded bomb which one dares not touch. But just as one can let a bomb blow up in a safe place and so render it harmless, one can release murderous feelings safely in the therapeutic setting. I help patients do this all the time. Once released, the underlying anger can be handled rationally.

The third step to liberation is self-possession. This means that the individual knows what he feels; he is in touch with himself. He also has the ability to express himself appropriately to further his best interests. He is in command of himself. Gone are the unconscious controls stemming from the fear of being himself.

Gone are the guilt and the shame about who he is and what he feels. Gone are the muscular tensions in his body that block his self-expression and limit his self-awareness. In their place is self-acceptance and the freedom to be.

In the course of this book I will explain how one arrives at this stage through the therapeutic process. That involves an analytic investigation of the individual's past so that one understands why and how the self was lost or damaged. Since the experiences of childhood that created the person's problems and difficulties are registered and structured in the body, reading the body can provide basic information about the past. This knowledge, plus what is gained from the interpretation of dreams, the analysis of behavior and the discussions with the therapist, must be connected by the patient with what he feels, and with the sense of his body. Only in that way are mind and body integrated so that the person is whole.

Therapy is a voyage in self-discovery. It is not quick, not easy and not without its fears. It may actually take a lifetime in some cases, but its reward is the feeling that one's life has not been in vain. One can find the meaning of life in the deep experience of joy.

CHAPTER 3

CRYING:
THE RELEASING
EMOTION

The chronic muscular tensions that stifle and imprison the spirit develop in childhood out of the necessity to control the expression of strong emotions: fear, sadness, anger and sexual passion. Of course, these controls are not always effective since feeling is the life of the body and at times that life will assert itself despite the individual's attempts at control. The neurotic individual's control can break down in a hysterical outburst of crying and screaming, in a wild rage or in sexual acting-out. Such actions are not ego syntonic and do nothing to resolve the conflict between the need to express one's feelings and the fear of expression. Until that conflict is resolved the person is not free to be himself. Originally the fear of expression was related to a fear of the consequences that might follow such expression, but while it is true that the fear still persists in the adult, it is now an irrational fear. For example, expressing anger in a therapy session about one's treatment as a child will certainly not result in any punishment or other serious consequence. The fear is of the feelings themselves. They are seen as threatening and dangerous. Many people have a murderous anger in them because their spirit was broken as a child and they are unconsciously afraid that if they lose control they might kill someone. In the forty-eight years that I have worked with patients, encouraging them to feel and express their anger, not one has ever gone out of control to where they turned on me or broke anything in my office. They beat the bed with their fists or with a tennis racquet as hard as they could, but they knew what they were doing and were in conscious control of their actions. The fact is that few of my patients could get angry enough to where their eyes blazed with their fury. In a succeeding chapter I will describe how I work with my patients to help them feel their anger: It is not enough to know that one *is* angry; one has to *feel* it. The same thing is true for fear, sadness, love or sexual passion.

One can't feel an emotion unless one can express it in a gesture, a look, in the tone of one's voice or through some bodily movement. This is because the feeling is the perception of the movement or impulse. I draw a distinction between an emotional expression and a hysterical outburst. In the latter, the ego (which is the organ of perception) is not connected to the action, with the result that the action is not perceived as an emotion. It is not uncommon to see someone go into a rage and deny that he is angry. When I screamed in Reich's office, I was not aware of feeling frightened. I regularly see patients whose bodies show all the signs of fear—wide-open eyes, raised shoulders, restricted breathing—but who deny feeling any fear. The bodies of these individuals manifest fear but expression, which is an active and conscious process, is absent. This disconnection is especially common with sadness.

I believe that people are more afraid of their sadness than any other emotion. That may seem strange since sadness doesn't strike one as a threatening feeling, but the fear is connected to the depth of the sadness. In most patients it touches despair, and they are afraid, consciously or unconsciously, that if they let go of their effort to hold themselves up they will sink into the depths of despair without any hope of coming out. But if they don't allow themselves to feel their despair, they will spend their lives struggling to stay up without any sense of security and certainly without any good feeling. By going into the despair one can find that it stems from the childhood situation and is not relevant to the life of the adult. Adult situations can trigger feelings of despair because they are connected to similar situations in early childhood which gave rise to feelings of despair. An adult can replace a lost love defect but a child cannot through its own effort replace a parent. Of course if one uses all his energy to hold one's self up or to present a positive facade of denial, one will never find the security, peace and joy that life offers. The fact is that some patients cannot cry and most cannot cry deeply, which prevents them from feeling their sorrow and blocks them from ever finding joy. To help them one must understand the pattern of tension that blocks their expression and know the body techniques that will help them get through the blocks.

A person begins therapy because he is hurting in one way or another. He may be anxious, depressed, confused, frustrated or just plain unhappy about his life. He is hoping the therapy will enable him to change this condition, improve the way he functions in the world and find some good feelings—maybe even some joy. He is hurting because he has been hurt. While some new patients are aware that their childhood was unhappy, that they were frightened and lonely, most believe that their misery is the result of some weakness or defect in their personality. They look to therapy to help them overcome their weaknesses, in effect, to make them stronger. This picture of the average patient has changed considerably in recent years as people have become more sophisticated about therapy and have learned that emotional problems stem from childhood traumas. Many now want to learn about their past to understand why they feel and behave as they do, but they want to use this knowledge to change, so that life will be more fulfilling. Unfortunately, that can only be done to a very small degree, because the effects of the past are structured in the body and beyond the reach of the will or conscious mind. Deep and meaningful change can occur only through the surrender to the body, through the reliving of the past emotionally. The first step in this process is crying.

Crying is an acceptance of the reality of both the present and the past. When we cry we feel or sense our sadness and we realize how much we hurt and how badly we have been hurt. If a patient says to me, "I have nothing to cry about," as some have done, I can only answer, "Then why are you here?" Every patient has something to cry about, as do most people in our culture. Certainly, the lack of joy in our lives is something to cry about. Some patients have said, "I've cried a lot but it does no good." That is not true. Crying will not change the outside world. It will not bring love nor acclaim. But it will change the inner world. It will release the tension and the pain. That can be understood if one observes what happens to a baby when it starts to cry.

A baby cries when it is in distress. Its crying is a call for the mother to remove the cause of the distress. Distress causes the baby's body to contract and stiffen, which is the body's natural reaction to pain and discomfort. A baby's body reacts more intensely because it is more alive, more sensitive and softer. It also lacks the ego capacity to tolerate pain. Unable to support the tension, it begins to quiver. Its jaw puckers up. A moment later its body convulses with deep sobbing. The sobs are convulsions that run through the body in an attempt to release the tension caused by the distress. A baby will continue to cry as long as the distress continues or until it is exhausted. When its energy is depleted and it can no longer cry, it will fall asleep to protect its life. Crying has a similar effect on children who are overtired and cannot quiet down. This state of tension makes them restless and fretful. Often, this will result in the mother becoming angry and perhaps even hitting the child, which will cause it to cry deeply. The crying has two effects: It releases the tension, relaxing the child's body, and it allows the breathing to become deeper and fuller. Generally, the child will then fall asleep. I don't recommend hitting a child, because that is a hostile act. One could use a sharp command to shock the child out of its hysterical state if that is necessary. I simply want to point out that crying serves to release an individual from a state of tension.

There is a common belief that a good cry can make one feel better. A "good cry" is one that is deep and continuous enough to release a significant part of the tension resulting from some emotional distress. Such crying takes the form of sobbing, which is accompanied by rhythmic waves that flow through the body. This is the only kind of crying that will release the pain, the hurt feelings and the muscular tension of an emotional crisis or trauma. The crying of tears is also a tension-releasing mechanism for the eyes and, to some extent, for the body, since it softens with the feeling of sadness. Eyes become frozen with fear, contracted by pain and dulled by sorrow. The flow of tears is a melting and softening process like the thawing of ice in the spring. Eyes that do not cry become hard, brittle and dry, which can impair their visual function. Tearing is a very human action. No other animal actually cries with tears. Such crying in humans reflects their ability to see the sadness, pain or distress in another person or creature. That is why most people *cry* when seeing a sad movie, but very rarely does a sad movie make us *sob*. I believe, therefore, that the ability to shed tears, to cry, is the basis of the ability to feel compassion, whereas when we sob, we express our own deep sorrow and pain.

Sobbing is not the only form of vocal expression stemming

from feelings of sadness, sorrow or distress. If the pain of the distress is intense and seemingly unending, the crying may take the form of wailing. Wailing is a more high-pitched, continuous sound. It expresses a very deep hurt, one that is felt in the heart. Such a hurt would result from the death of a loved person which is why wailing is a typical reaction of women who have lost a loved one. Men's voices do not naturally make the high pitched wailing sound which women's voices can. Another sound that belongs to the category of crying is moaning. The moan, in contrast to the wail, is a low-pitched sound. One moans from a pain that seems unremitting and one that is of long standing. There is an element of resignation in the moan which is absent from the wail or the sob. These sounds are associated with pain, distress, hurt and loss. They are sounds of sadness and sorrow, not of joy. Joy has its own range of vocal expression. The laugh, for example, is very much like a sob except that it has a positive note, an upbeat ending. There are screams of delight just as there are screams of torment. Thus one can sing the happiest tunes as well as the saddest.

It is important to recognize that the voice is the medium for the expression of many feelings. We can also express feelings through our actions, but that expression comes from a different place, namely, the muscular system of the body, which is the mechanism of action. A smile, an embrace, a blow, and a caress all express feeling. When a person does not sense the feeling of an action, it is because the action is mechanical and he is dissociated from it. The same thing is true of the voice. Many people speak in a dry, mechanical tone that expresses no feeling. Here, too, we are dealing with individuals who are dissociated from their body and have subjected it to the control of the ego. Many people, for example, cannot cry with sobs because this expression of feeling has been suppressed by chronic throat tensions. Others cannot feel or express anger. Such individuals are emotional cripples who cannot experience joy or any strong emotion. In my opinion a therapy which does not help an individual recover his natural function of self-expression has failed. In this chapter we will explore the problems patients have with vocal expression.

Voice is the result of vibrations produced in the column of air as it passes by the vocal cords. Variations in the sound are created by changes in the diameter of the opening in the throat and larynx, by which air chambers in the head are used to create the resonance, and by the amount of air. The human voice has a very wide range of expression that corresponds to the range of emotion the person can feel. Not only can the voice express all the emotions mentioned earlier, but it can vary the intensity of the sound to correspond to the intensity of the feeling. The voice is one of the main avenues for the expression of feeling and, therefore, for self-expression. Any limitation on the voice constitutes a limitation on self-expression and represents a diminution of the sense of self. The saying that one has "no voice in his affairs" refers to this connection. Since all patients suffer from some decrease in their self-esteem or in their sense of themselves as having the right to "speak up," it is important to work with the voice in a therapy which seeks to enhance the self.

Many children go through painful and frightening experiences which cause them to lose their voices. Renée described herself as an incest survivor. She was sexually abused by her father, with penetration before she was six years old. The pain of the experience was so great that she left her body and lived in her head. The term "left her body" means that she had no feeling awareness of it. She would look at her feet and be surprised that they were attached to her body. To a lesser degree this was also true of her arms. Despite this severe degree of dissociation she was a survivor. She had been married twice, had raised three children and was able to support herself. But the men she married abused her physically and sexually, and she could not resist for many years. With the support of incest groups she gained the courage to enter therapy and begin the difficult task of regaining her self.

In the course of one session she remarked how difficult it was for her to speak out. She said, "When I have to express myself by saying, 'How dare you. Who do you think you are?' I feel choked. I become afraid that I will be choked to death for speaking out. About three years ago I had a flashback of a scene from my childhood. I was standing by a door, holding the knob and preparing to leave the room. I was about nine years old. I was facing my father and I remember saying to him, 'If you don't stop, I'll tell Mommy.' He grabbed me around my throat and shook me. I felt I was going to die. But he never touched me after that."

During the therapy I would have Renée kick and scream "Leave me alone." She could only do it with my encouragement and strong support, and for a minute or so she would let herself go in an hysterical outburst. But when it ended, she withdrew, curling up on the corner of the bed like a very frightened child and whimpering with fear. Then she would slowly come out of it more connected to her body and her self. She also did the grounding exercises mentioned in the preceding chapter, which furthered her sense of connectedness. Her normal speaking voice was controlled, young-sounding, and light. It was a voice that came from her head with very little body resonance and, therefore, little feeling. To use her voice as a form of self-assertion was extremely difficult for her. Screaming "leave me alone" was a body voice which came from feeling, but it was not connected to her ego or her head. That is the nature of a hysterical reaction. It denoted a split in the personality. When Renée spoke from her head, there was no body feeling. When she screamed hysterically there was no ego identification.

Screaming, by its very nature, always has an element of hysteria in it because it is an uncontrolled expression. One can yell with control but one can't scream with control. To scream is to "blow your top" or "blow your lid," which means that the ego is overwhelmed by the emotional outburst. It is a cathartic reaction in that it serves to release tension. In this respect it functions like the safety valve on a steam engine that will blow off when the pressure becomes too great. A person will generally scream when the pain or stress of any situation becomes intolerable. If one is unable to scream under these conditions one could actually lose one's mind and go crazy. Crying—that is, sobbing—also serves to reduce tension and release strain, but one generally cries when the trauma or injury has ended. The scream, on the other hand, is an attempt to avert the trauma, or at least limit the attack. It is an aggressive expression, while crying is the body's attempt to release the pain following an injury. Screaming and crying are involuntary reactions, although in most cases the person can initiate or stop the reaction. Sometimes the reaction gets out of control and the individual screams or cries hysterically, seemingly unable to stop. But one always stops when the fury is spent. We have a taboo in this culture against uncontrolled behavior because we are frightened by it. It is also regarded as a sign of weakness of character, of infantilism. And in a sense, when one screams or cries, one does revert to a more infantile type of behavior. But such regression may be necessary to protect the organism from the destructive effect of suppressing one's feelings.

The ability to let go of control at an appropriate time and place is a sign of maturity and self-possession. But the question could be asked: If one consciously decides to let go and give in to the body and its feelings, is one really out of control? What control does an individual have who is terrified of screaming and so blocked from crying that he cannot express these feelings? The ability to let go of ego-control also includes the ability to maintain or reestablish that control when it is advisable or necessary. When a patient lets go in a bioenergetic exercise, kicking and screaming and seemingly out of control, he is generally fully aware of what is happening and can stop it at will. It is much like riding a horse: If the rider is afraid to "let go" to the animal, if he tries to control the horse's every movement, he will soon find that he has no control at all. The person who is so frightened of letting go of control isn't in control at all. He is controlled by his fear. As one learns to "let go" to strong feelings through the voice and movement, one loses the fear of surrendering to the self.

As we know, babies can scream so loudly that they can be heard at a great distance. They can also cry freely. It is amazing how powerful a baby's voice can be. When my son was an infant he suffered from colic. When an attack hit him, he would scream so loudly that he could be heard two blocks away. Only my parrot can scream louder than that. When that parrot screams, it is as if its whole body becomes a voice box. The vibrations of its voice are so strong that no tension can hold against it—some voices are known to shatter glass. One of my problems was an inability to use my voice freely. My eyes could tear very easily but sobbing was very difficult for me. More than twenty-five years after my therapy with Reich, I had an insight which explained why my voice wasn't free. During a Bioenergetic Analysis workshop, two participants who were themselves therapists offered to work with me. I was hesitant but I let myself go. One of the women worked on my legs and feet as I lay on the floor, massaging them to release some tension. (There has always been considerable tension in my legs; my calves hurt under pressure.) The other woman worked on my tight neck. Suddenly I felt a very sharp pain in the front of my neck as if a knife had been drawn across my throat. I knew immediately that what I experienced was the physical aspect of what my mother had done to me psychologically. She had cut my throat. On a very deep level

I was afraid to speak up to her as a child, and fear made it difficult for me to speak up to others when I was an adult. Considerable work on this problem over the years has greatly ameliorated that condition.

Another patient, whom I will call Margaret, told me of a recurrent dream in which a pillow is over her face and she has the feeling of being suffocated and dying. Margaret was another survivor, but just barely. She could function seemingly normally, but she was always in a deep state of anxiety and fear that made her life almost intolerable. Margaret was terrified of her mother, even in her late forties, when I saw her in therapy. She described her mother as a cold, hard, controlling woman. Margaret's way of survival was to withdraw emotionally while she carried on her life with almost no feeling. She existed largely in her brain.

Margaret had considerable difficulty giving in to the sadness she felt. If she started to cry, she became nauseous and had to stop. This went on for a considerable time before the nausea let up and she could begin to cry. But her sobbing did not flow. It sounded more like broken wails—vain attempts to open her throat and let out her pain. Her speaking voice was thin, flat and heady. She would talk quickly with no emotional expression. What she said made sense but had no feeling.

To help her I put some pressure with my fingers on the sides of her throat to ease the tension while she tried to scream. Her throat was so tight screaming was almost impossible. But our work over the past year had loosened her somewhat. To my surprise, instead of choking she opened her throat and let out a full sound. When it ended, she said to me, "I never heard that voice before." It was the voice of the child that had been buried in her body all those years.

Children are born innocent with no inhibitions or guilt about their feelings. For many, there are feelings of joy in this early state of bliss. As I look at young children between one and two years of age and make eye contact with them, I see their eyes light up and a look of pleasure appear on their faces. Invariably, they turn away in shyness or embarrassment, but in a few minutes or less they look again at me to recapture the excitement and pleasure of that contact. They turn away again but not for long. It's a game the child could play for a long time, while I stop because the cares and responsibilities of adult life intrude and make me leave.

I have also seen adults light up when such eye contact is made, but it is so tentative and so quickly ended that one can sense their embarrassment and guilt. However, there are very many whose eyes do not and cannot light up because the internal fires of the spirit, which we call passion, have been severely dampened. One sees this in the darkness of the eyes, in the sadness of the facial expression, in the grimness of the jaw and in the tightness of the body. They lost the capacity for joy early in their childhood when their innocence was smashed and their freedom destroyed. Martha is a case in point. She was a fifty-one-year-old woman, the mother of three grown children and recently divorced when she came to see me because, as she said, her life had no meaning. What she meant was that there was no joy in her life and very little pleasure. She said that she had always felt anxious and had believed that it was the normal condition.

I was struck at our first meeting by the darkness around her eyes. There was no brightness in her eyes themselves and at no time in the consultation did her eyes light up, even momentarily. She was a small but well-proportioned woman. Her manner was lively and, despite a grim expression in her mouth and jaw, she did not act depressed. After many years of marriage, during which she served her husband faithfully, he left her for another woman. Martha took the divorce stoically and continued her empty life until she realized that she needed help.

Martha knew that she was frightened. She had never been

able to stand up to her husband. The divorce left her in a very insecure position, and she had never supported herself before. Now that she was approaching menopause, she felt hopeless. But she didn't admit to feeling hopeless and she didn't cry. Martha was a survivor, one of a very large number of people in our time who carry on and survive, but with no joy in their lives.

I have heard a number of people say with pride, "I am a survivor." I can appreciate that feeling if one has lived through life-threatening situations like the Nazi concentration camps. But the statement also has meaning for the present and the future. In effect, the individual is saying, "I can take it. I can survive conditions that others would succumb to. I can withstand hostile or destructive 'attacks'." If one is geared to survival, one neither anticipates, nor can one respond to, joy. Can one expect a knight in armor to dance a waltz? An attitude that prepares one to meet disaster does not predispose one to the enjoyment of life. This is not to say that these individuals who identify themselves as survivors do not want pleasure. But wanting pleasure and being open to it are two different things. If survival is the focus of life, one is not open to pleasure. If one is armored against a possible attack, one is not open to love. Opening to life makes such persons feel too vulnerable and their fear closes them up again.

Martha was the youngest of three children, all girls. The atmosphere in her home was one of potential violence. Her parents fought all the time, mostly over money. She recalled one incident when she was five: Her parents were yelling at each other in the living room when suddenly her father kicked over the coffee table and was about to smash the china closet when her sisters stopped him. Relating this incident, she did not mention that she was frightened. I believe she didn't feel her fear because she was in shock. She did remark that "it was very scary."

In this atmosphere Martha withdrew and closed off. She said that she used to hide and play alone under the dining room table where she was covered by the table cloth. She considered it her house. But it was not a sanctuary. She never felt free from fear. "I lived in a constant state of anxiety as to what could happen," she remarked. "There was no joy or lightness in my home. The mood was heavy, like drudgery. It was a heavy sadness."

In her perpetual state of distress Martha had no understanding, sympathy or support from either parent. When, at six, she had to go to school, it was a terrifying experience. Her mother brought her to the school and, when she turned to go, Martha cried and begged her not to leave, but her mother ignored her pleas and left. Martha said that she spent the day in a corner crying and crying.

It struck me that Martha had spent her childhood under a dark, threatening cloud. Survival dictated that she had to pull herself together and go out into the world since she couldn't spend her life under a table. She married a man she didn't love as soon as she left high school. She had learned a way of coping; if she did what was expected of her, she would not be hurt. She would be a good little girl. Her husband turned out to be very much like her father, an angry, violent man. But she knew she could survive.

Coming to therapy meant that Martha wanted more than survival from her life. To get more meant that she had to make a major change in her attitude toward life, and that would require more than a simple decision. If her attitude had enabled her to survive, giving it up meant putting her survival on the line. While her present situation was such that there was no real threat to her survival, dropping her defensive posture and opening herself to life evoked the feelings of vulnerability and danger that she knew as a child. Despite her 51 years and sophistication she was, underneath, still that frightened little girl. She still suffered from anxiety, had feelings of distress and was insecure.

If the road to joy is through the surrender to the self—that is, to one's feelings—the first step in the therapeutic process is to sense and express one's sadness. To have spent 51 years just surviving is a sad story. To express that sadness, one needs to cry, but while Martha could see the sadness in her face, it was very difficult for her to cry. Lying over the bioenergetic stool, Martha could feel the distress in her body. Getting her to use her voice in a sustained sound resulted in a little crying and the words, "Oh God, oh God."

"Oh God" is a person's deepest and most spontaneous appeal for help. We all say it sometimes when we reach a point where we feel that the pressure or pain is too much. It is not the cry of a survivor who feels that he must not break down under any condition. We utter it when we feel we cannot take it anymore, when we feel "it" is too much. The amazing thing about these words is that if they are uttered with any feeling, they could easily lead to crying. The word "God," with its two consonants —g and d—on either side of the short vowel, resembles the sound of a sob. When people break into deep crying—that is, sobbing—they will often say quite spontaneously, "Oh God, oh God!"

When Martha uttered these words, I suggested that she tell God what she felt. However one thinks of God, whether as a religious deity or a supernatural force, one could pour out one's heart to God without fear of humiliation or rejection. It is easier to tell God "I hurt" than it is to say it to another person when you know they might not want to hear it. Martha's reaction to my suggestion to talk to God was as follows:

"You're mean. You're not good. You don't love me," and then, "I don't know what I feel—I feel, I feel, I feel, I don't know." Not to know what one feels denotes a terrible confusion, a lack of self-awareness, a very inadequate sense of self. One has to feel bad in that condition. I asked her, "Don't you feel terrible?"

"Yes," she answered, "I don't feel good. I'm not happy. I feel

sad. I feel very, very sad." But she didn't cry. Instead she remarked, "I can't breathe."

After this, she said, "I am angry, too." Her voice was very
small. It sounded like that of a child. When I pointed it out to
her, she said, "It is very hard for me to express anything. I get
that way with people, too. I can't talk. I keep thinking, 'Children
should be seen, not heard.' "

If the crying is choked off, one can't breathe. One has choked
off the flow of air by constricting the throat. Martha's throat was
severely constricted, which accounted also for the small childlike
voice. That constriction is at the base of her inability to breathe
deeply and easily.

Her difficulty in breathing became more apparent in the
course of another exercise. This exercise is a second step in the
expression of feeling. It involves kicking, which is an expression
of protest. We all have something to kick about. We have all
been hurt in some way which we feel we did not deserve. We
have a right to protest, to ask "why?" or to kick.

I had Martha do this kicking exercise while lying on the bed.
The directions were to kick the bed fast and strong with straight
legs and at the same time say, "why?" With my encouragement
she kicked rapidly and let her voice rise to a scream. She let
herself go a little wild for a couple of minutes. When it was over,
she laughed and said, "I feel good." Then the anxiety returned
and she remarked, "I felt that if I let go and lost control, I'd lose
my life." It was the first time she mentioned a fear for her life.
I did not explore further during this session, but it was clear to
me that all her life she had been afraid for her life. If she
screamed hysterically, she would be killed. She would be strangled. The constriction in her throat was directly related to the
fear of being strangled. It was as if someone had a hand on her
throat in a threatening gesture.

The ability to cry out and to speak out is the basis for an

individual's sense that he has a voice in his own affairs. Prisoners and slaves have no voice in their affairs and are not free people. But children can also fall into this category if they have been so frightened that they cannot make a loud sound. If not actually slaves, such children learn to be submissive and quiet as a technique of survival. That technique generally persists into adult life and cannot be surrendered until the person has the experience that yelling and screaming will not result in punishment. On the other hand there are individuals for whom screaming (hysterical behavior) is almost a way of life. I believe that both patterns develop in families where violence and potential violence is typical of parental behavior. If the child is not terrified, it can identify with the parents and adopt their pattern of behavior. On the other hand if the child feels threatened and frightened enough by this behavior, it will withdraw into itself and become quiet and submissive.

All infants react to any form of distress by crying, which is considered to be a call for the mother to remove the cause of the distress. All mammalian infants will call for the mother when in distress. But the crying of the human infant is more than just a call for help. Even when the mother responds, the child's crying may continue for some time. Furthermore, it is not a single note or call which may be repeated, but a continuous broken sound connected to the rhythm of breathing. It is the sound of sobbing which adults also make when in distress. Sobbing can also be seen as a call for help, but it has a deeper significance for adults and for children. It expresses the emotion of sadness or sorrow. Sadness is also associated with the flow of tears, but in many cases when a person is sobbing deeply there are no tears. In other cases the tears flow but there is no sobbing.

Because sound and feeling are so closely connected, we have learned how to control our voice so as not to reveal our feelings.

We can speak in a flat, unemotional tone which denies feeling or we can sound "up"—that is, raise the pitch of the voice—to hide the fact that we feel down. This regulation of our voice is exercised largely through the control of respiration. If we breathe freely and fully, our voice will naturally reflect our feelings. By breathing shallowly, we stay on the surface of our feelings where we can consciously control the quality of our vocal expression. One way to get a patient in contact with his deeper feelings is to deepen his breathing. The technique I use is very simple: The patient lies over the bioenergetic stool breathing easily. Then I ask him to make a sound and to sustain it as long as possible. Some will make a loud but short sound. This could indicate that they would like to open their voice but cannot. Others might make a soft sound which implies that they do not feel entitled to express themselves strongly. In both cases, the person is still in control. I then suggest that they use every effort to continue the sound. This involves forcing the expiration. When they do this, their control begins to break down. Toward the end of the breath one can hear the sound of a cry, a moan or a note of agony. As one forces the sound, the vibrations extend deeper into the body. When they reach the pelvis, one can hear and see that the patient is on the verge of crying. Repeating the exercise again and again, encouraging the patient to hear the tone of the sound, will often induce the crying.

In most cases, however, I have found it necessary to direct the patient to break the voice into repeated gruntlike sounds—"ugh, ugh, ugh." This sound will send vibrations through the body just as the sobbing does. Most patients do not feel it as a sobbing sound, which in fact it is because they are producing the sound mechanically. But if I keep them making the sound, especially at a more rapid rate, it will become involuntary and the patient will feel it as true crying. It's like priming a pump. The deliberate action induces a feeling which turns the movement into an expressive act. The word "God" when intoned has a similar quality, and if repeated quickly can also end in crying.

Crying is an acceptance of our human nature, that is, of the fact that we have been expelled from our earthly paradise and live with the consciousness of pain, suffering and struggle. But it seems we have no right to complain, for, having eaten the fruit of the tree of knowledge, we have become like gods, knowing right from wrong, good from evil. That knowledge is the cross we bear, the self-consciousness that robs us of spontaneity and innocence. But we bear the cross proudly because it makes us feel special, that we *alone* are God's creatures even though *we* are the ones who violated God's first injunction. Man has gained other knowledge, too, which has now given him the power to destroy the Earth, his true Garden of Eden.

Self-consciousness is both man's curse and his glory. It's a curse because it robs him of joy, the joy of blissful ignorance. It is his glory because it offers him a knowledge of joy as ecstasy. An animal experiences pain and pleasure, sorrow and joy, but it has no knowledge of these states. To know joy is to know sorrow, even when it is not immediately present in one's life. It is the knowledge that we will lose our loved ones and even our own life. If we reject this knowledge we reject our true humanity and the possibility of knowing joy. But this knowing is not a matter of words but of feeling. To know and to feel that human life has a tragic aspect, that sorrow is inevitable, allows one to experience a joy that is transcendent. We have been hurt and we will be hurt again, but we will also be loved and honored—honored for being fully a human being.

To live life as a full human being requires the ability to cry freely and deeply. If one can cry freely and deeply, there is neither confusion nor despair nor torment in it. Our tears and our sobs wash us clean, renewing our spirit so that we may rejoice again. William James writes, "The stone wall inside of him has fallen, the hardness of his heart has broken down. . . . Especially if we weep! For it then is as if our tears broke through an inveterate dam—leaving us washed and soft of heart and open to every nobler leading."[1]

But crying does not produce miracles. We are not changed that much by one good cry. The issue is to be able to cry freely and easily. I had broken down twice in the course of my therapy with Reich and each time a seeming miracle occurred. But that crying, deep as it was, came as a result of external pressure. When other problems arose my jaw tightened as I tensed to contend with them. I had come close to failure but in the end I didn't fail. I knew that I could not cry easily. On one occasion during my work with Pierrakos, my associate in the early days of Bioenergetic Analysis, I asked him to apply pressure to my jaw. As I was lying on the bed he placed his two fists against the sides of my jaws and pressed. It was painful but I didn't cry. Then, as he continued the pressure, I said spontaneously, "Dear God, please let me cry" and I broke down into deep sobs. When I got up Pierrakos told me that my head was surrounded with a brilliant glow.

But even that experience, great as it was, needed to be repeated. The object of the therapy was not to get me to cry (that would be provoking it, which has to be done sometimes) but helping me recover my ability to cry freely and easily. That happened many years later as I began to work with my patients to help them cry. If I kept a sound going long enough as I lay on the stool, it would break into sobbing sounds with which I could identify and to which I could surrender. To keep that surrender in place against the pressure of a character determined not to

[1] James, William, *The Varieties of Religious Experience* (New York: The Medusa Library, 1906), p. 262.

give in, I had to cry regularly. At times I have done some crying every day. If someone would ask me, "What are you sad about?" I would answer, "Me, you and the rest of the world." When people look deep into my eyes, they report seeing a sadness in them. But my eyes still have the capacity to light up when I make a warm contact with another person's eye.

When patients tell me that they have cried enough, I point out that crying is like the rain from Heaven which fecundates the earth. Should we ever say, "Enough rain, we don't need any more?" We may not need a deluge, but we certainly need a gentle rain regularly to keep our planet green and our souls clean.

Both sadness and joy stem from sensations in the belly. We noted in the preceding chapter that the orgasm reflex occurs when the respiratory wave flows freely into the pelvis. In this surrender to the body there is a sense of freedom and excitement which produces the feeling of joy. Fear of letting go to the sexual excitement blocks the wave so that it does not reach the pelvis, producing a feeling of sadness. If the sadness can be expressed in crying, the tension will be released, freedom and fullness reestablished and a good feeling in the body restored. The involvement of the belly in both sadness and joy is reflected in the remarks "a good belly cry" and "a good belly laugh." Both result in the person feeling good. Of course, the individual who can breathe deeply into his belly and can cry or laugh with that depth of feeling feels good about himself and has no need of therapy.

If crying and laughing are similar in their energetic and convulsive patterns, can we not heal ourselves not only through crying but through laughter as well, as Norman Cousins did? Both actions have a cathartic effect in releasing a state of tension. But laughter is ineffective and meaningless in freeing an individual from his suppressed sadness or despair. It may lift him temporarily out of the sadness but he will fall back into it when he stops laughing. It is much easier for a person to laugh than to cry. One learns early in life that laughter brings people closer whereas crying may push them away. "Laugh and the world laughs with you, cry and you cry alone." Many people have difficulty responding to another's crying because it touches their own pain and sadness, which they are struggling to deny. But fair weather friends are not reliable. A true friend is one who can share your pain, and he can do this because he has accepted his own pain and sorrow.

In many people laughter is a cover-up. It can have a value in sustaining one's spirit in a crisis, but in these cases it is not the deep belly laughter of true enjoyment. In the course of working with the voice, a patient will sometimes break into spontaneous laughter instead of crying. The situation, however, is not appropriate for laughter. The person is in therapy because of serious problems in his life which he admittedly has difficulty facing. Laughter in this situation must be seen as a resistance to surrender, a denial of the reality of one's feelings. When I have pointed this out to the patient, his response is, "But I don't feel sad." Rather than confront his resistance, I will join him, laugh with him and encourage him to laugh even harder. In most cases, as the laughter deepens the patient will burst into sobs and feel the sadness that lies under the surface of his consciousness. Following such crying the person feels a great sense of relief and freedom.

Women find it easier to cry—that is, to sob—than men. I believe this is cultural, in that men or boys are shamed if they cry. But the facility a woman has to cry is also related to her body structure, which, generally, is softer than that of men; I attribute the woman's greater longevity to her softness. As a rule men are more rigid in their bodies; they do not break down easily. But when this rigidity is unconscious, an habitual posture or characterological attitude, it amounts to a lack of responsiveness to life and represents, therefore, a loss of spontaneity and vitality. Dead men don't cry. I believe that a man who can cry lives longer. Crying protects the heart.[2] It is the only way to release the pain of heartbreak, of the loss of love. Life is a fluid process which becomes completely frozen in death and partially frozen in states of rigidity, which are states of tension. Crying is a thawing out. The convulsive sobs of crying are like the break-up of the ice in a spring thaw. Tears are the resulting flow.

Most of us, however, have been too deeply and too badly hurt. We carry too much pain in our bodies to allow us to surrender to the self. Our sadness reaches despair which we must deny in the interest of survival. Our fear can be paralyzing so that we can only function by suppressing and denying the fear. We cut off feeling by tensing the body and restricting our breathing, but in doing so we also cut off the possibility of joy. To help my patients I point out to them that the despair is not of the present but of the past. The fear does not stem from a present threat but a past one. True, the feelings of despair and fear are in the present but only because we have embalmed the past in our bodies. The past lives on in the tension. Releasing the tension allows us to step free from the past.

But the tension can only be released if one expresses the feeling that is contained in the tension. Relaxing techniques only help temporarily. As soon as a life situation arises which can evoke the blocked feeling, the musculature contracts again to control the feeling. But discharging the feeling in a hysterical outburst, though it may be cathartic, does not produce a lasting release either. It is important to understand the dynamics of self-expression if we are to help our patients become free. The ego is an integral part of self-expression as much as the body is. Mind and body must be integrated in any expression of feeling for it to represent an affirmation of the self. Thus crying or even

[2] Lowen, Alexander, *Love, Sex and Your Heart*, (New York: Macmillan Publishing Company, 1988; Arkana, 1994).

screaming is not therapeutic unless one knows why one is crying and can express it with words. I have seen patients yield to crying when breathing over the stool and then say to me, "I don't know what I am crying about."

If sound carries the feeling, words express the image or idea which gives the feeling meaning. Bioenergetic Analysis is a therapeutic mind-and-body technique which works with feelings and ideas, with sounds and with words. Most patients, when they yield to deep crying, will say and often repeat the words "Oh, God," which I described as an involuntary appeal for help. If the crying sound is an appeal for help, words communicate that appeal on an adult level. When a person expresses a feeling in words as well as in sound or action, his ego is identified with the feeling. Often a patient will scream spontaneously in the course of a cathartic release and then say, "I heard myself screaming but I wasn't connected to it." Putting words to the feeling helps establish the connection.

As in the session with Martha described earlier in this chapter, once people say "Oh, God" in the course of their crying, I suggest that they tell God what they feel. In some cases they say, "I don't feel anything" or "I don't know what I feel." In such a case I might say to the patient, "Do you feel sad?" "Yes," they answer. "Well, then, tell God you feel sad." They will say, "I feel sad." Often the words are flat and I will ask them, "How sad?" They will always answer, "Very sad," which is the truth of their self. If I can get them to use the words with some feeling, their crying deepens. Some patients open up easily and say "I hurt, I am in pain" or other statements which express the image and idea associated with their sadness and crying. The more they can express in words what they are crying about, the more integrated they are. Mind and body then work together to provide a stronger sense of self.

Sometimes I get a very negative response to my suggestion

that the patient tell God what he feels. One patient said, angrily, "Fuck you, God. You were never there for me. You didn't care. I hate you." This was a woman who had been brought up in a religious home and had attended a religious school. When I made an attempt to question her feelings, she said it was how she felt and who she was. Her father was perverted in relation to women and sexuality. He was sexually interested in my patient, touching her and looking at her in a seductive way. While he saw all females as sluts and told dirty jokes at the dinner table, he made negative, humiliating remarks about anyone else's sexual expression. He expected his daughter to be an angel but he saw her as a whore. The remarks about God allowed my patient to sense more clearly the hypocrisy in her family and to feel how it had made her bitter and disgusted with men.

God represented her father, which suggests that before she had become aware of genital feelings, at about the age of three to four, she had adored her father, as most little girls do. Her subsequent experiences with him were perceived as a betrayal of her love. Her anger against him was beyond words, for she felt that her spirit had been killed. All of these feelings were projected upon me as God, as the therapist, as a substitute father and as a man. I shall postpone to a later chapter the issues of resistance and transference which are so critical in every therapy, and which can only be dealt with by words, but if these words are to have any value, the patient must be in touch with his feelings. A patient who doesn't feel his sadness and cannot cry cannot be reached with words.

One reason for the focus on the body in Bioenergetic Analysis is that words alone are rarely strong enough to evoke suppressed feelings. The suppression of feeling is the work of the ego, which observes, censors and controls our actions and behavior. Words are its voice just as sound is the voice of the body. One can dissimulate with words. It is quite difficult to do that with sound. One can recognize a note of falseness when a sound does not truly express a feeling. It is an axiom in Bioenergetic Analysis that the body doesn't lie. Unfortunately, most people are blind to the body's expression, having been taught very early to believe the words one hears rather than what one senses. But some young children still retain an innocence which allows them to trust what they see. We recognize that the moral of the story "The Emperor's Clothes" is that only the innocent can see the truth. Children have not yet learned the sophisticated art of playing with words to hide one's feelings. I'll never forget the man who consulted me in my early years as a therapist, saying, "I know that I was in love with my mother." It was as if he had said, "Doctor, see if you can tell me something new." I didn't take up the challenge and the therapy never got off the ground. I should have said, "What you don't know is how sick you really are."

The same blindness is shown by people who, when I point out their need to cry, answer, "I have no difficulty crying. I have cried a lot." The latter part of the answer may be true, but the former is not. Their difficulty lies in the inability to cry deeply enough to reach the bottom of their sadness. Their crying is like water over the dam which never empties their lake of tears. The fact that they need help dealing with life denotes the existence of distress and a lack of joy, which is something to cry about. Many people are taught as children that crying is only acceptable when they are destroyed, not when they are hurt and in pain. Children who cry when they've been hit have been told, "Stop crying or I'll really give you something to cry about." And in some cases, children have been given a double measure of punishment to stop them from crying. Children are shamed into not crying: Little boys don't cry, only girls cry. And adults are discouraged from crying: One should be brave. Crying is a sign of weakness, and so on. I have found that the ability to cry is a sign of strength. Because women cry more easily than men, they are clearly the stronger sex.

When a person cries, each sob is a pulse of life that goes through the body. One can actually see the pulse traverse the body. When the pulse reaches the pelvis, it causes a forward movement—the person crying can actually feel the pulse strike the pelvic floor as it passes down through the inner tube of the body. That's a "letting down to the bottom." Such deep crying is as rare as is deep breathing. There is, however, another dimension to crying which is the amplitude of the wave. That is expressed in "size" of the sound. A full sound means a wide-open mouth, throat, chest and abdomen. The degree of opening determines how open one is to life—to taking it in and letting it out. When we describe a patient as being "closed in on themselves," it is literally true of the body's apertures. The lips could be tight, the jaw clamped, the throat constricted, the chest held rigidly, the belly flattened and the buttocks pulled in. In these individuals, the eyes are also narrowed.

Therapy is a process of opening to life—a physical operation as well as a psychological one. It is reflected in bright eyes, a warm smile, a gracious manner and an open heart. But opening the heart without opening the passageways through which the feeling of love flows to the world is an empty gesture. It's like opening the safe deposit boxes of a vault but keeping the vault door locked. I always start the therapeutic program by helping a person open his voice (to speak out with) and his eyes (to see with) before he opens his heart. But this process of opening up is neither quick nor easy. It's like learning to walk. The patient tests the ground with each step he takes. He has to learn to trust himself and then to trust life again. And, like a child who has to fall over and over while learning to walk, the patient will also fall, sense his fear and feel his impotence, but as he gets up and goes on to try again and again, he will grow in faith, in confidence, in wisdom and in joy.

Crying deeply may result in a breakthrough in which one senses freedom and feels the joy of it. Such breakthroughs are like the sun shining through breaks in the clouds—not a sign that the storm is fully over but an indication that the end of the storm is near. Each breakthrough makes one stronger and more open to life, more able to surrender to the body.

In the next chapter, I will discuss the resistances to crying. They are big and deeply structured in the personality. And they cannot be abandoned without understanding that they developed as a means of survival.

CHAPTER 4

THE RESISTANCE
TO CRYING

I Won't Break Down

In the preceding chapter I pointed out that most people have a need to cry to discharge the pain and sadness of their lives. Crying, or sobbing, releases the tension that holds these painful feelings locked in the body. It is the natural response to having been hurt physically or psychologically. All trauma is a shock to the organism, which causes it to freeze or contract, to stop breathing and to close up like a clam. Crying is the process of unfreezing, decontracting and opening up to life. After the convulsions, breathing is relaxed and deep. It restores to an individual the full use of his voice and refreshes his soul as a good rain refreshes and recharges the earth. People who cannot cry are frozen, their bodies are tight and their breathing is severely restricted. No person will recover his full potential for being if he cannot cry. To recover that full potential the crying has to come from the pit of the belly. That is not easy for most people whose breathing and crying are restricted by moderate to severe diaphragmatic tension.

In this chapter I would like to discuss the psychological resistances to crying that parallel the physical blocks. In our culture crying is regarded by most people as a sign of weakness. Even in situations where crying is a natural response, as in the case of the death of a loved person, the bereaved is often admonished to be strong and not to give way to the sadness. Yielding to one's feelings can meet with strong disapproval. Giving in to one's feelings does represent a loss of ego control, but if the surrender of ego control is not appropriate in this case, when would it be? Crying is seen not only as a sign of weakness but also of immaturity—childish or infantile. Children are often mocked for crying: Big boys don't cry. It is, of course, true that crying is associated with helplessness. In a situation of danger, it may be necessary not to give in to a feeling of helplessness and cry, but at therapy, the patient is not threatened by his helplessness except on an ego level.

Many men have the mistaken idea that it is manly not to cry. John had a similar belief. He consulted me because he was severely depressed. He said that when he didn't have to go to work he would lie in bed all day, unable to mobilize himself. John was a good-looking young man in his early thirties who aspired to be an actor. In his acting class he had heard about Bioenergetic Analysis and how it worked with the body to help people to get more in touch with themselves and increase their ability to express feelings. He had been in therapy with a psychologist who he believed was helping him and wanted to continue that relationship while he also worked with me. I had no objection since I could only see him every other week for one hour.

John looked "manly." He had a strong-looking, muscular body which he attributed to a program of weight-lifting when he was younger. The outstanding feature of his appearance was a swagger that he accented by wearing cowboy boots. He was aware that his appearance reflected a strong narcissistic element in his personality, but he regarded this as an asset. His breathing was very shallow, as I could see when he would lie over the stool, and I encouraged him to work strongly with the exercises I've described to deepen his breathing, develop some vibration in his body and express some feeling. He did the exercises but without much feeling. He would smile at me as if to say, "I don't think this will work." Nevertheless, he always felt better after the sessions and it was my hope that he would come to realize their value. At this time John was living at home with his mother, although he had lived on his own for a number of years. John had a younger brother who was married and seemingly doing well. His father had died when he was young, putting him in the position of being the man of the family.

John's depression stemmed from the fact that while he was set up to be the man of the family, he was undermined by a domineering mother with whom he was emotionally involved. He recognized that there had been sexual feelings between them. I knew that his depression would lift if I could get him to cry but we never arrived at that place. However, he did relate an incident from his childhood which shed light upon his resistance to crying. He said that when he was six years old, his mother locked him in the bathroom and beat up on him all day. She finally stopped when he was broken and crying heavily. He never did break down and cry during the short period of our work together. Then, one day, he said to me, "You won't get to me. I won't cry." His depression didn't lift and against the advice of his other therapist he signed himself into a hospital. I did not see him again after that.

I'm sure that it was very difficult physically for John to cry, but, in addition, he had a strong conscious will not to cry. This will was part of his ego defense system. When he said, "You can't reach me," he also meant that you would not break him down. His mother had done that once, but while she may have made him cry, his inner core had toughened until he could resist her with the strength of steel. It has to be appreciated that this resistance saved his integrity. Had she broken it, he would have become schizophrenic. Since his resistance had enabled him to survive, he was not about to surrender it. It also froze him into an attitude of defiance which left him with no energy or freedom for any pleasure or creative action. Little wonder he was depressed. My experience with John made me realize how strongly some people resist crying.

I generally start the body work by having the patient lie over the bioenergetic stool and breathe. This allows me to observe the breathing and note the quality of the respiratory wave. The position is slightly stressful, which actually forces the patient to breathe more deeply. In no case is the patient's breathing as full or free as it should be. To deepen the breathing I ask the patient to make a loud sound and to sustain it as long as possible. In almost all cases the sound is too short and flat. Holding in the breath is the means whereby one holds against giving in to the body and its feelings. This "holding" is unconscious. The new patient generally believes that if he made an effort he could let the air out more fully and sustain the sound longer. He is encouraged to keep trying to prolong the sound. Continuing the sound allows the respiratory wave to reach the belly where feelings are. If the sound is continued long enough, one will generally hear a note of sadness come into the voice. Sometimes the voice will break and some sobbing sounds will be uttered. Occasionally the patient will break into some deep sobbing. It is never deep enough in the early stage of therapy to release the pain and suffering, but this experience provides an opportunity to discuss the patient's attitude toward the expression of sadness.

It is amazing how many people come to therapy presenting problems that are debilitating yet who deny any feeling of sadness. This is especially true of depressed patients who, having suppressed their emotions, are emotionally deadened. If depressed people could cry, their depression would lift since they would feel alive again. Sadness, however, is not the only emotion that is suppressed. Anger is equally suppressed. People can get irritated, go into a rage, even become violent, but feeling and expressing a clear emotion like sadness or anger is very difficult. Expressions of irritation or even of rage are not intended to effect any significant change in the person's situation. They are minor releases to relieve the tension of frustration and are much like "blowing off steam." Once the tension is released, the person feels better but the situation has not changed. Anger, on the other hand, doesn't subside until the painful situation is cleared up. The same is true of sadness. If one feels deeply sad, he will institute some changes in his life. To know that one is sad or angry helps, but it is not enough. To feel the sadness or anger one must be able to express it. Babies and young children can do this easily when they are hurt. How does this natural reaction become blocked in individuals?

Joan was a married woman in her thirties whose several years of therapy had done little to relieve her feelings of frustration and depression. Looking at her body, I could understand those feelings. Her head was small and held rigidly above the body. Her face was tight with a bitter expression. Her body was soft, harmonious but boyish and immature in its form. The split between her head and body indicated that her ego was not identified with her body. The body's boyish aspect denoted a desire to deny her femininity. Unable to accept her true nature or to escape it fully, she was a tormented and frustrated woman. It is not surprising that she was depressed. In several previous sessions we had worked on her inability to express any deep feeling. Through the grounding and the breathing exercise over the stool, she got her legs to vibrate, giving her some sensation in her body, but no emotion broke through. Some of her frustration and bitterness were expressed through kicking, during which she yelled, "Leave me alone."

At her next session Joan described an experience she had had a week prior in a bioenergetic study group. She remarked that other people in the group were crying. Some said that they had sexual feelings. Joan added, "My body was vibrating, my pelvis was moving, but I didn't feel anything. I don't trust people. I don't give in. I don't surrender to anything. I guess I don't trust myself." This was a very clear statement of the nature of her problem. She would not surrender to her body. In some way surrender to her body threatened her survival. She had to dissociate her consciousness from her body, which created the split between the two. Therapy had to help her understand what had happened and why.

Lying over the stool and breathing, Joan felt the tension in her back, which represented her rigidity, her unyielding quality. She would not bend nor break. She felt the pain and said, "It hurts but I won't cry. Only sissies cry. I can endure it." This was followed by, "You're not going to break me. Goddamn it. You're not going to break me. I'm not going to give in. You'll break the chair before you break me. It hurts." A little later she said, "You're trying to make me give in or give up but I'll go to hell before I do." Joan realized that the issue was not between her and me; she knew that the conflict was between her mother and herself. She said, "There was a power struggle between us. I had to have some part of myself. She possessed most of me. I did what she wanted. I gave her everything but my feelings. If I surrendered those I would become her thing, her plaything. When I didn't give her what she wanted, it drove her crazy."

Mike related a story similar in many respects to John's (which I described earlier), except that he did not suffer from depression. He had achieved some standing in his profession but felt that his life was without meaning or pleasure. His body was severely split: its upper half did not match the lower half. He had wide, straight shoulders and a big chest. His waist was narrow and severely constricted, and the lower half of his body was small and un-derdeveloped. Pointing to his wide shoulders, I remarked, "You are well-prepared to shoulder some heavy responsibilities." He smiled and said, "I've been carrying people all my life." What I did not tell Mike was that he struck me as a broken man. When he spoke, his voice was weak and without feeling.

The story he related was that he was the oldest of three chil-dren with a mother he described as crazy, as a woman afraid of life. He said, "She beat me every way she could to break me. I wasn't permitted to cry. I had to take it." He described his father as unavailable, either working or drinking. But where John had developed a very strong resistance to his mother, Mike had sub-mitted. He became her little man, serving her as his father didn't. This submission resulted in the loss of much of his manhood and self. John's resistance enabled him to preserve some sense of man-liness, which he attempted to project in his swagger, his cowboy boots and his pretension to be an actor. Mike, on the other hand, had given up his resistance. It was his way of surviving. Another important difference was that while John wouldn't cry, Mike couldn't; he didn't have the voice.

Breathing and vocalizing over the bioenergetic stool helped Mike's voice become a little stronger, but not to the point where he could cry. In contrast to John or Joan, Mike had an uncon-scious resistance to crying. His ego was identified with his ability to "take it" and with his role of shouldering responsibilities for others. Crying would be an admission of failure and an accep-tance, on an emotional level, of the emptiness and sadness of his personal life. However, his coming to me for help signified some willingness to face this issue.

It is essential for all patients to protest the way they were treated as children. Without a strong protest, one cannot break free from the horror of the past. I had Mike lie on the bed and kick strongly, voicing the words, "I can't take it anymore." With my encouragement he let go, kicking wildly, and screaming, "I can't take it anymore." Then he added, "Oh, God! It's so sad, so much pain." And he began to cry a little.

One is at a loss to understand maternal behavior which could have such a devastating effect on a child. What possessed John's mother to beat him so mercilessly? What strange force drove her against her own deepest feelings to break her son, to destroy his spirit? Why did Joan's mother need to possess her body and soul? The psychological, physical, and sexual abuse of children is common and well-known today. All of my patients have suffered some ill treatment from one or both parents. What I find particularly distressing is the cruelty inflicted on children by parents who were themselves victims of cruelty. Some had been in Nazi concentration camps. Such behavior seems to reflect a law of human nature: Do unto others what has been done to you. Parents will raise their children as they were raised. Many patients have told me that their parents were treated as harshly as they were. I am sure that John's mother was beaten by her father and I am also sure that she felt justified in her attack upon her son. It would take an enlightened parent to stop the progress of this destructive acting-out on children. What is required for such enlightenment will be discussed in the next chapter.

The survivor is generally characterized by a strong will which has enabled him to survive. In many cases it has also enabled him to be fairly successful in the world. I have worked with a number of people who had risen to important positions in the professional and business world through the use of strategies based on the will to survive. One of these strategies is the denial of feeling and the reliance upon a sharp, calculating intellect. This might seem a great asset in a world where feelings are a handicap, where the dominant values are power, money or prestige and where the competition for success is intense. In such an environment one subordinates feeling to the drive to succeed. But while some do gain some success in terms of money, power and prestige, their lives are emotionally empty: no intimate, fulfilling relationship, no real pleasure in their work and no joy. One sees the latter in the dullness of their eyes, in the lack of a high energy charge in their movements. Many suffer from some depression and most of them complain of chronic fatigue and tiredness. The basic dynamic in these individuals is a dissociation from the body. One individual who consulted me said of herself, "I was identified with my job. I was a management consultant for a large firm. I felt a sense of power and I had a lot of responsibility in my work which gave me a feeling of worth, but I worked too hard, too much and I became depressed."

Another woman tells a similar story: "After college, I went off in search of a career. Diligently, I worked my way up the corporate ladder. Having achieved an executive position, I benefited from working with professionals worldwide. Everything was fine until, at the age of thirty-six, the one and only intimate relationship I allowed ended in the man's abandoning me. For the first time in my life, I suffered from depression." That was only the beginning of the breakdown of her narcissistic "second nature." She left her job to start a new career in the helping professions, which was a positive move, but six months later she had a serious automobile accident. She recovered but was left in a severe state of anxiety that manifested itself in an intestinal disorder known as irritable bowel syndrome, the symptoms of which were cramps and diarrhea. This syndrome stems from a state of chronic tension in the lower intestines which I believe is related to fear. She described the effect of this disorder on her personality as follows: "I could always control my mind; I was now forced to recognize my helplessness to control my body. It was a horrifying and frightening experience. During this period, I literally went to bed every night in the "fetal" position because I was so frightened of what was happening in my body. For the first time in my life I could not deny nor conceal my vulnerability."

For all survivors, the surrender to the body is strongly resisted because it brings up the most painful and frightening feelings. If vulnerability is the issue, how dare one cry deeply, since the feeling associated with that crying would be one of helplessness. Ann had lost her mother at the dependent and helpless age of five. Following her mother's death she was raised by a number of surrogate mothers who abused her both emotionally and physically. Unfortunately, during this time of pain and loss, fear and helplessness, her father was critical of her. He blamed her for not being pretty like her mother, smart like her mother, sweet like her mother, and so on. His basic attitude to life was "only the strong survive." Ann learned that one must not express emotional pain and did what every other survivor learns to do— dissociate from the body and withdraw into the head.

Cut off from the body, one doesn't feel vulnerable. By identifying the self with the ego, one also gains the illusion of power. Since the will is the instrument of the ego, one truly believes "where there's a will, there's a way" or "one can do whatever one wills." This is true as long as the body has the energy to support the ego's directive. But all the willpower in the world is no help to a person who lacks the energy to implement the will. Healthy individuals do not operate in terms of willpower except in an emergency. Normal actions are motivated by feelings rather than by the will. One doesn't need willpower to do what one wants to do. There is no need to use the will when one has a strong desire. Desire itself is an energetic charge which activates an impulse leading to actions that are free and generally fulfilling. An impulse is a flowing force from the core of the body to the surface, where it motivates the musculature for action. The will, on the other hand, is a driving force that stems from the ego— the head—to act counter to the body's natural impulses. Thus, when one is afraid, the natural impulse is to run away from the threatening situation. However, this may not always be the best action. One cannot always escape a danger by running. Confronting the threat may be the wiser course, but this is difficult to do when one is frightened and there is an impulse to run. In such situations mobilizing the will to counter the fear is a positive action.

The situation described above is commonly faced by children whose parents are abusive and threatening. Some young children actually try to run away from such homes but their attempts at escape are hopeless. The child must accept the situation and yield to the parent but, at the same time, he must find some way to maintain its integrity. His submission must not be total, his will must not be broken. He stiffens and rigidifies the body so as not to break down, which is an action mediated by the ego through the will. The child sets its face in an expression of determination not to surrender, lose control or be overcome by fear. The so-prevalent chronic jaw tension stems directly from this need to be in control. Once the will is mobilized by the chronic rigidity and tension in the body it becomes a driving force for power and leads to a way of life in which the struggle for power is the underlying theme of the person's life. In this situation, crying is seen as a breakdown of the will and surrender is impossible. Life is lived as if the body were in a state of continual emergency. Of course, no joy is possible.

The Surrender of the Will: Despair

People come to therapy because they need to change some aspects of their behavior and personality. On a conscious level they want to change, but at the same time they are resistant to changing. That resistance stems in large part from their desire to be in control of the process of change. Submitting to the therapeutic process involves giving up this control, which the patient sees as submission to the therapist. This gives rise to feelings of vulnerability in the patient and to the idea that he will be misunderstood and abused as he was when he was a child and helpless in the family situation. Because of this background, the patient sees the therapist as having power over him which he must oppose to maintain his own integrity. Therapy often degenerates into a power struggle, which in reality is nothing more than the patient's struggle to avoid surrendering.

The idea of surrendering is frightening to most people. "Letting go" or "giving in" to the body and the self sounds more acceptable, but people do not know what that involves; in practice it turns out to be just as frightening. Neurotic patterns of behavior have developed as means of survival, and even though they now prove to be counterproductive in adult life, the individual clings to them as if to life itself. These patterns have become so ingrained that the person experiences them as part of his nature. True, it is second nature—the first nature was that of the innocent and open child—but that first nature was lost and seems irretrievable. By adulthood one has lived so long with his second nature that it feels comfortable, like an old pair of shoes. Still, when a person comes to therapy it is a tacit admission that this second nature has failed in important respects. But that doesn't mean that the patient is ready to surrender it. The change that he is looking for is to make his character or second nature work successfully. He is open to learning better ways of coping and acting but he is not prepared to surrender his strategy of survival.

This attitude in a patient is known as resistance. Sometimes it makes its appearance in the early stages of the therapy when the patient expresses a distrust of the therapist or questions his competence. Personally, I welcome a clear statement by a patient of his distrust. Having been hurt as a child by those whom he trusted, he would be naive to place his trust in a stranger about whom he knows little. Therapeutic competence is not guaranteed by diplomas or popularity. No therapist can change a patient any more than the patient can change himself. Therapeutic change is a process of growth and integration as a result of what the patient learns and experiences through the therapeutic process. The best judge of that is the patient himself. Unfortunately, most patients do not trust their own perceptions and feelings—which is part of their character problem. Being desperate, many are willing to turn over control to the therapist in the illusion that he can change them. The surrender I am talking about is to the self, not to another person. One follows the suggestions of a therapist, one does not submit to him.

The process of therapy starts with the consultation. We sit face to face and the patient tells me about himself, his problems and his background. While he is talking I have a chance to study him, that is, to note how he holds himself, the tone of his voice, the expression on his face, the look in his eyes, and so on. I will ask about his present life situation and about his childhood, seeking information that would explain his difficulties. I ask how he experiences his body, what muscular tensions he is aware of, what pains or illnesses he has experienced. Then I explain the body–mind connection, emphasizing the functional identity of the physical and the psychological. Many of the people who contact me are somewhat familiar with this approach, having read some of my books, experienced it or heard about it from other therapists. If the person is prepared and suitably dressed, I look at the body to see its pattern of tension. Generally I do this as the patient stands in front of a mirror so that I can point to and explain what I see. It is important for the patient to understand that the body has to change if the person is to change. Specifically, the tensions pointed out during the examination have to be understood and released if the person is to become free. To release these tensions, the individual has to sense their constricting effect, to understand how they control his present behavior and to learn how and why they developed. Finally, the impulses blocked by the tensions have to be expressed. At this point there is no talk of surrender. The focus is on awareness and understanding. The individual is furthering his identification with his body.

It is important to understand the depth of a patient's distress and difficulty. Mary was a young woman whom I first met when she was a participant in a professional workshop. When I looked at her body I saw a severe contraction in the area of her waist, which functionally split her body into separate halves. This meant that the wave of excitation associated with respiration did not pass into the lower part of her body. This split had two significant effects upon her personality: One, her heart feelings, located in the chest, were not connected with her sexual feelings, located in the pelvis. This disturbance seriously affected her relationship with men. Two, her body showed a deep sense of insecurity resulting from the lack of feeling in the lower half of the body, which undermined its ability to function as a solid base. I pointed this out to Mary and informed her that it could change if she worked on her problems bioenergetically, that is, both psychologically and physically. Mary later entered into therapy with me because, as she said, I was the only therapist who understood the depth of her problem. Others with whom she worked psychologically saw her as together, competent and successful. She was a rather competent therapist herself, successful in her practice and seemingly in a good relationship with her husband. But it was a good relationship only because she was submissive. She was able to put up a great front that deceived others but confused her. So many individuals seem normal to a superficial view, but when one looks at the body carefully, one sees the truth of the being. The body does not lie, but one has to be able to read its expression if one wants to know its truth.

Mary worked with me for several years. Her case is reported at greater length in a later chapter. As she became stronger and developed a greater sense of self she left her husband and experienced joy for the first time in her adult life.

Not every patient who consults me wants to hear the truth about themselves. Some narcissistic individuals are not open to learning the truth, which makes it almost impossible to work with them. I do not expect my patients to accept what I see but to be open to hearing it. They will learn the truth as they experience themselves on a body level. In the beginning, however, it is important to develop a working relationship with a patient. The best basis for such a relationship is the patient's sense that he is understood, that he is seen as a person who is struggling to find some fulfillment. Throughout his life he has been told that he needed to make a greater effort—to change one or another behavior pattern to feel good. If his fears were recognized he was advised that he could overcome them. He's been led to believe that his difficulties are only in his mind. He, too, can now see that they are also in his body and that working with body and mind in an integrated way can be more effective than a verbal therapy alone. The breathing and expressive exercises that I introduce him to generally have a very positive effect, giving him more energy and lifting his spirits. While these early experiences will not produce significant changes in personality, they are important in helping establish a positive relationship between us and constructing a solid base of understanding which will support the hard work that has to be done to free the patient from his hang-up.

Ego defenses are not purely psychological. If they were, it would be easier to surrender them. Most patients recognize that their defenses are handicaps, that the situation which gave rise to them no longer exists. However, the problem is that the defenses are structured into the body where their function is to suppress feeling. They are walls to contain and control frightening impulses. An individual cannot be robbed of the joy of life without feeling a murderous rage. How does one handle such an impulse in a civilized society? One does not tear down the walls of a prison housing dangerous criminals until one has found a way to defuse their hostility. It is a subject I will examine in the next chapter. But we also erect walls to hide behind, to protect us from being hurt, to hold back our sea of sorrow. Unfortunately, these walls also imprison us.

Patients don't let themselves cry because they are afraid of the depth of their sadness which, in most cases, borders on or amounts to despair. As one patient said, "If I start to cry, I may never stop." I have no hesitation in saying that most people harbor a despair of ever finding a true love, of ever being happy or joyful. When one of my patients told her mother that she was unhappy, that she wanted some happiness, her mother answered, "Happiness is not what life is about. It is about doing what one should." But without some feeling of joy, life is empty, frightening and painful. It is the pain of a hunger for connection that is as unbearable as the pain of a hunger for food. It is understandable that patients are reluctant to descend into this hell. But to deny it, to deaden one's self to the longing and the pain, is to accept a living death.

Deadening oneself may promote survival but the pain is not eliminated. It will surface from time to time as a purely physical pain in the form of chronic tension in some part of the body, making the person miserable. While it is still an emotional pain, one can diminish it through crying and surrender. The difference between a purely physical pain and emotional pain is that the former is localized and affects a limited area of the body, while emotional pain—also in the body—is generalized. Headache is a pain localized in the head, a toothache is limited to the area of the tooth, and a pain in the neck affects only the neck. In contrast, the pain of loneliness is felt throughout the body. Emotional pain stems from the body's contraction in response to the loss or disruption of a loving connection. Such experiences can be heartbreaking, especially when they happen to a child and are connected with a sense of rejection and betrayal.[1] Since the pain feels life-threatening to the child, survival demands that the experience, together with its pain and fear, be suppressed. Suppression is achieved by numbing the body through rigidification or by dissociating from the pain. Both procedures cut off feeling, leading to a sense of loneliness and emptiness. These conditions become painful when an impulse to open up and reach out arises and is blocked by the fear of rejection. Since these impulses cannot be completely suppressed as long as one is alive—they are the essence of the living process—the individual is in a struggle with his own nature, that is, with his body and its feelings. Actually, the struggle is between the ego, with its defense against rejection and betrayal, and the body, with its imprisoned heart. The tension which this conflict sets up in the body is experienced as pain. Surrendering to one's nature and allowing the impulse full and free expression immediately reduces the pain and results in the pleasurable feeling of fullness and freedom.

Since emotional pain represents a conflict between an impulse and the fear of its expression, it can be eliminated by suppressing the impulse fully or removing the fear that is blocking the full expression of the impulse. A patient named Julia recently complained to me about her lack of good feeling after some months of therapy. We were talking about her sexual relationship with her husband, whom she experienced as being needy. His ad-

[1] Lowen, A., *Love, Sex and Your Heart.* Op cit.

vances left her cold, yet she enjoyed other aspects of the marriage. I had always encouraged Julia to be true to herself and her feelings and I had supported her in not submitting sexually when she had no desire. This support had allowed her to make some significant progress, but she was still in conflict. "I am afraid to tell you what I feel," she remarked. "I'm afraid to say I don't love my husband because you will tell me to leave him. If I say 'I don't feel I am getting through in therapy,' you will tell me to stop." This was the same conflict she had had with her mother, who had told her, as I reported earlier, that happiness (joy) is not what life is about.

In her mother's view, life was about being there for others. Julia explained that she was regarded as "the special one" by her mother. "She said I was her real child, her silken one. She needed me and I had to be there for her. That's how I lost myself." Julia understood that when she cut off her feelings, it left an empty space in her personality that her mother took over. This surrender *of* the self—not *to* the self—made her constantly feel lonely, empty, unfulfilled and sad. "But," she added, "I am very reluctant to go into that place, although I know it's true. It hurts so much I jump right back into my head." Julia withdrew from her belly, where she would feel the sadness of the loss of her self, but this very act of withdrawing was a surrender of the self. I might add that the withdrawal upward also cut off much of her sexual feeling, which contributed greatly to her sexual conflict with her husband.

All feelings arise from bodily processes and should be understood in terms of those processes. Many of these feelings stem from and reflect experiences of the past. Julia's sadness reflected the painful feeling of the loss of her bodily self. When she said, "It hurts so much," she was talking about the conflict between the need to cry and the holding against it. The pain of that conflict can be agonizing. She remarked, "I feel like I'm on a rack being tortured. I can't stand it yet I feel I have to stand it. If I don't my parents will leave me." That fear was transferred to me. If she didn't improve, I would leave her. Even though Julia knew that her fear was irrational, it was a real feeling that could only be discharged through the expression of her anger and not by an act of will. She felt a lot better after this discussion because she had expressed her fear and realized that it stemmed from a conflict in her childhood. It was related to the present only through the holding back of its expression.

Almost all patients have some fear of abandonment stemming from experiences in childhood. In most cases that fear, which amounts to panic, is not consciously perceived because it is blocked by the rigidity of the chest wall. By keeping normal respiration to a minimum, one stays above the feeling of panic, but such shallow breathing also cuts off all feeling, leaving one empty and unfulfilled. On the other hand, to feel the panic is extremely frightening and painful, but one can get beyond it by breathing deeply. The feeling of panic is directly connected to the sensation of not being able to get one's breath. The reason one has difficulty breathing is that the muscles of the chest wall have become contracted by fear—fear of abandonment. One gets caught in a vicious circle: The fear of rejection or abandonment ------> difficulty in breathing ------> shallow breathing ------> panic when one breathes deeply, thus bringing forth the fear. The individual is forced to live on the surface, that is, unemotionally. On this level, he can keep above the underlying feeling of panic, but such living, although seemingly safe, is relatively dead. This mechanism keeps the fear of abandonment alive. If one breathes through the fear, one will cry deeply and feel the fear of abandonment as a carryover from the past. The deep crying also releases the pain of the loss of love, as I pointed out earlier. Thus, by surrendering to the body and crying deeply, one passes through the fear and the pain into the calm water of peace where one will know the joy of being free.

Julia's case enables us to understand the pain of loneliness, which is the physical side of the fear of being alone. That fear creates a need for other people and for activities to distract the individual from feeling alone. Since the distraction is only temporary, the person is faced again and again with the fear of being alone. The fear isn't rational but it is real. Of course, not everyone is afraid of being alone. People can be alone if they can be with themselves. But if one doesn't have a strong and secure sense of self, being alone feels empty. The feeling of loneliness stems from a sense of inner emptiness which, as in Julia's case, is a consequence of the cutting off of feeling.

One can't be lonely if one is emotionally alive. Even when alone, one feels part of life, of nature and of the universe. Many people prefer being alone to the hassles which seem part of today's relationships. Others accept being alone because they have not found a person with whom they would want to share their life. Such people are not lonely; they are not in pain and do not feel empty. Without the ability to be alone one is a needy person looking for someone on the outside to fill the inner emptiness. There is no joy in such a life because it is lived on the survival level, namely, "I can't live without you."

The irrationality behind the fear of being alone is evident in the mistaken remark, "If I accept being alone, then I will always be alone." The fear overlooks the fact that the human being is a social animal who wants to live with others and intimately with one other. We are drawn toward each other because contact increases our aliveness. But this positive effect is absent when one individual becomes a drag on the other through depression or neediness. Some neurotic individuals need to be needed, but arrangements based on need sooner or later create resentments which easily turn into deep hostility. Both the person who needs and the person who is needed lose their freedom and the possibility of joy in the relationship.

The only healthy relationship in which needing and being needed is inherent in the situation is the one between a parent and a child. The parent who fulfills the child's need fulfills his own as well. The child who is not fulfilled becomes a needy person in adulthood who feels that he needs someone to "be there" for him. The feeling is genuine although it doesn't belong to the present and cannot be fulfilled in the present. If one tries to respond to this need, one infantilizes the individual without helping him. The individual's present-day need is to function fully as an adult, for this is the only level on which he can be fulfilled. The blocks, both psychological and physical, which impede adult functioning must be removed. That is done by reliving the past with the understanding of the present. By breathing and crying deeply one can feel the pain of the loss of support and love in childhood. One can then accept the loss as from the past and be free to fulfill his being in the present. A child could not do that because the love and support of the parent was essential to his life. Survival demanded that the loss be denied. The child had to believe that he could regain the love through some effort on his part. He would submit to the parent's demands even to the point of sacrificing himself, as Julia did. But if this sacrifice insures survival, it guarantees unfulfillment, emptiness and loneliness. The despair is buried in the pit of the belly and never released.

Any attempt to overcome the loss and the pain of the past through the will doesn't work. Its failure perpetuates the despair. Accepting the despair but realizing that it does not represent the present allows one to pass through it. The principle is exemplified in the story of the farmer whose horse was stolen and who now stands guard at the barn door with a shotgun against the theft of the horse. Like all neurotics, the farmer, by denying the reality of the past, is condemned to live it again. The surrender to the body constitutes an acceptance of the reality of the present. While the principle is clear, its application is not easy. Surrender requires more than a conscious decision because the resistance is largely unconscious, structured in the body. The tight, determined jaw can be momentarily softened, but it returns to its set, determined position as soon as consciousness is withdrawn. It is an old familiar habit which has become so much a part of the personality that one feels awkward without it. But if one is committed to a surrender of the tight, determined attitude of the jaw, one will find that the new, relaxed position of the jaw feels right and that the old, tight position is now uncomfortable. But this changeover takes considerable time and work, because surrendering one's determined way of being will affect one's entire behavior in the world. It amounts to a real change in one's life style, from one of doing to one of being, from one of toughness to one of softness. Also, the letting go of chronic tension can evoke considerable pain, because when one attempts to stretch tight muscles, it hurts. The pain is held in the tight musculature but is not felt. Tight muscles have to be stretched before they will let go.

In many individuals the tension in the jaw is associated with a retracted jaw rather than one thrust forward in an aggressive attitude. Both positions block the surrender by immobilizing the jaw so that its free movement is restricted. Thus, while the forward thrusting jaw expresses the attitude, "I won't let go," the retracted jaw says, "I can't let go." Freeing the jaw from its locked position requires considerable work and elicits pain. But the pain of stretching tight muscles disappears when the tension is released, whereas the pain of temporomandibular joint disease, which is caused by chronic jaw tension, increases with time. Persons suffering from this condition cannot fully open their mouths, restricting both their breathing and their voice.

Chronic tension in the muscles of the jaw is not an isolated phenomenon. A tight jaw is always accompanied by tight throat muscles that restrict the ability of the person to voice feelings. A tight throat makes it extremely difficult to cry or scream. I use special breathing exercises with my patients to help them release this tension, but it is slow work. Even if the individual breaks through and cries deeply, the release is not permanent. Muscles are elastic and quickly resume their accustomed state. One has to cry again and again, each time a little deeper and freer, until crying is as easy as walking. One has to practice screaming again and again until it feels as natural as talking. A good place to practice screaming is in an automobile on a highway with the windows rolled up. One can scream one's head off and nobody hears.

The surrender of the ego also requires that the muscles at the back of the neck be soft, especially those connecting the head to the neck. Tension in these muscles is so common in our culture because we are all operating from our heads and the fear of losing one's head is great. "Don't lose your head" is one of the basic injunctions in our society. But if we don't let go of ego-control, how can we surrender to the body and to life? How can we fall in love if we don't let go of our heads? People who are always in their heads have difficulty falling in love or falling asleep. This tension in the muscles at the base of the skull where the head joins the neck is responsible for all tension headaches. It is also responsible for many eye problems because this tension circles the head at the back of the eyes. And it also spreads down through the muscles at the back of the neck, making the free rotation of the head difficult. This stiffness at the back of the neck represents an obstinate, stubborn attitude. Individuals with this attitude are described as stiff-necked. Such stiffness persisting over the years will also create an arthritic condition in the neck vertebrae, which can be very painful.

These tensions cannot be released through massage or manipulation alone. They represent characterological attitudes developed early in life to deal with distressing situations by controlling and cutting off feeling. These attitudes must be understood both historically and in their present-day function. In addition, the feelings they contain must be expressed. The major feeling controlled by these tensions is sadness as expressed in the statement, "he broke down and cried." Through analyzing the resistance to crying and by getting the patient "to break down and cry," much of the tension can be released. Another part of the tension can be discharged through screaming. In a scream, a tremendous energetic charge flows upward through the head to be discharged in the scream. In a scream, one "blows his top," one "loses his head." The scream is a safety valve which permits the safe discharge of a large pent-up force.

The way a person carries his head is significant in terms of his characterological attitude. Here are two cases which illustrate this idea. Larry was a business entrepreneur who felt that he had not been able to realize his potential for living. He had had much analytic therapy, but it had not greatly changed him. A strong, alert man, he sat facing me as we talked, with his head thrust forward. As we discussed his problem I became aware that he was well-defended. He would accept my observations easily but then he would explain his behavior logically and nothing would change. Physically, he had a tight chest which greatly restricted his breathing and blocked his crying. On one occasion, working with the bioenergetic stool, he almost cried, but it turned into laughter which continued for more than fifteen minutes. The laughter was a defense against crying. I believe that the first break into his defenses occurred when I suddenly understood his head position. Looking at his head held far forward, I realized that Larry was "ahead of himself." This meant anticipating every situation before it arose and thinking, calculating, planning how to deal with it. Such an attitude gave him a competitive advantage in his business, but it robbed him of the spontaneity and freedom that could make life joyful and fulfilling. He grasped my point quickly and it opened the way for some progress in the therapy.

The second case involved a man close to sixty who consulted me for his hypertension. Robert was a big man, successful in his profession and content in his marriage. Still, something was amiss in his personality, for he had developed a serious degree of hypertension. Looking at Robert's body, I could see that he held himself up. His chest was inflated, his shoulders were raised and he held his head up and tilted sideways and backward as if he were looking over people rather than at them. The upper half of his body was larger than the lower half. The simple interpretation of this posture was that Robert held himself above the common people as if he were a superior being. When I showed him how he held his head, he remarked that his grandfather had held his head the same way. Robert grew up in Northern Italy where his family felt themselves to be important because they were related to a count. Consciously, Robert didn't think he was superior, but that feeling could be read in his bodily expression. When I pointed it out to him he acknowledged the feeling.

In addition to the hypertension, Robert also suffered from lower back pain related to a band of tension around the waist that blocked the downward flow of excitation and so kept his blood pressure up. He also held himself above the lower half of his body, which represented his animal nature and which is the common ground of all humanity. We can see ourselves as superior only through the functions of the head, not through those of the pelvis.

To reduce his blood pressure Robert had to "let down," which is to surrender. He needed to cry because he was not fulfilled, not joyful, despite his seeming success. He wore a perpetual smile on his face, concealing an underlying sadness. It was not easy for Robert to cry since it would necessitate giving up the facade of the superior man. Robert was willing to do this on the conscious level but it was not so easy to change his bodily attitude. Getting him to breathe deeply and make a continuous loud sound while lying over the stool brought him close to a sob. He became conscious of how tight his chest was and how difficult it was for him to breathe out fully. Then when he bent over into the grounding position, his legs began to vibrate, which made him realize how little feeling he had in them. Working again on the stool with breathing and sounding allowed some continuous sobs to break through. In the grounding position, again, the vibrations in the legs became stronger. I also had Robert do some kicking, which increased his ability to let go. When he stood up at the end of the session, he said he felt much more relaxed and closer to the ground, and his blood pressure was almost normal.

Robert accepted the need to do some of the bioenergetic exercises at home. He had a stool made which he used regularly to deepen his breathing and allow some of his sadness to come out. He also did the kicking regularly. Both helped him to feel more alive. This also reduced his blood pressure, but it didn't stay down. Robert was using the exercises to overcome his problem, not to deal with it. He lived in another country and so I only saw him infrequently. When his blood pressure failed to stay down despite these exercises, he consulted me again. This time I pointed out to him that he was holding himself up to deny that he was a broken man. The break was evident in his lower back, where a very strong band of tension about his pelvis cut off any feeling of passion in his lovemaking. He knew about the tension but he had not accepted the fact that it caused a break in his personality, a split-off from his full sexual nature. That problem cannot be healed by crying, although one will cry when one feels the crippling it causes. Feeling that pain and disability, one can only react with an intense, almost murderous anger. Robert had suppressed his anger as he had suppressed his sexuality. That suppression had to be lifted if Robert was to recover his full self. Anger is the healing emotion.

Most individuals have severe muscular tension in the upper back and shoulders. This tension is related to the suppression of anger and cannot be released until these impulses are allowed expression. The problem of suppressed anger will be dealt with in the next chapter.

There is a resistance to crying which stems from a deeper source than those discussed in the preceding section—namely, despair. I have heard many patients say that they resist the surrender to their sadness and crying because they are afraid it will never stop. That thought is irrational; one cannot cry forever. But the feeling underlying it is real. I have replied that, of course, it will stop. But this reassurance doesn't go very deep and the fear remains. The patients' sorrow is felt to be a bottomless pit which they can never come out of once they let themselves go into it. Another metaphor patients use to express their despair is the feeling that they would drown in their sorrow. The idea of "drowning in one's tears" is more than a metaphor. Many patients have complained at times that they felt the liquid in their throats when crying, which gave them a sensation of drowning. Not having experienced that sensation myself I can only guess that tears are flowing back down into the throat through the sinuses. But the feeling could also be the reexperiencing of a sensation of drowning which the person had lived through at some early period of his life. Children do swallow water when learning to swim, and sometimes choke, and can develop a fear of drowning. Another possible explanation is that the individual had swallowed some amniotic fluid when in the womb, which might lead to a sensation of drowning. The embryo does make respiratory movements in the womb when it experiences a temporary loss of oxygen due to a spasm of the uterine artery. These sensations and anxieties act to constrict the throat, with the result that both breathing and crying are restricted.

Aside from these physical factors, the resistance to deep crying has a strong psychological core in the fear of despair. Every person who comes to therapy struggles with a feeling of despair— the despair of never finding true love, feeling free, or realizing one's full self. Despair is a terrible feeling. It undermines one's will, weakens the desire to live and leads to depression. As a consequence, the person will do his utmost not to feel his despair, and to hold himself up above the pit. This effort takes considerable energy and does absolutely nothing to remove the despair. Sooner or later, as one's energy decreases, one slips into the despair, into depression, illness or even death. If a person wants to get well emotionally and physically, he needs to confront his despair, which means to feel it fully and to understand that it stems from experiences in childhood and has no direct connection with one's adult life. As long as a person is afraid to breathe deeply, there is no real possibility of fulfillment. One will have a sensation of emptiness in the pit of the belly regardless of the external conditions of one's life. Marriage, children, and success in the world will do nothing to fill this emptiness in the belly which is energetically related to the fear of feeling one's deep sadness or despair.

Despair is the opposite of joy, which in the adult is intimately connected with the fullness of sexual excitation and discharge. In most people sexual excitation and discharge is largely limited to the genital apparatus and does not involve the whole body. Sex is not consciously experienced as being an expression of love because the genital apparatus is not connected to the heart and its feelings. The separation of these two centers results from the inability of the respiratory wave to pass through the relative deadness and emptiness of the lower belly and pelvis due to the suppression of feeling in that part of the body. The result is that sex becomes a localized bonfire, not a passion that consumes the being and results in the experience of joyous fulfillment that can reach the height of ecstasy. The fear of despair blocks the full surrender to the body in crying, which is the only means of releasing the individual from his despair.

Despair is frequently transferred to the therapeutic situation. After an initial surge of hope resulting from the early break-through of feeling, therapeutic progress slows down and may even stop. Some patients express their sense of despair that therapy will never really work, while others plug on. This development is a sign that the patient is struggling to fulfill some ambition or to realize a dream. Both of these objectives have the aim of finding love, a special love which the patient was promised as a child but which he never obtained. It was an erotic love based on a special relationship or intimacy between the parent and the child and which made the child feel special to that parent. It had strong sexual elements which greatly excited the child but which, at the same time, robbed the child of his innocence and freedom. It was the forbidden fruit of adult sexual love, sensed but not possessed. Nevertheless the child will be imprinted by the overwhelming excitement of this attraction and will unconsciously devote his life to the attempt to fulfill an impossible dream.

The impossible dream is to be "the special one." It is a narcissistic attitude that will drive the individual to prove his superiority in one way or another—actually, in the way the seductive parent desires. But this special love is not a deep connection between two individuals since it is based on appearances, not feeling. If love is a special relationship between two individuals it is because love is a special feeling. It is the love that makes a relationship special, not the specialness of the individuals that makes it loving. Such relationships necessarily fail to be satisfying and lasting, and these individuals will come to therapy feeling some despair but hoping that therapy will enable them to realize their dream of being seen and loved as special.

That desire is transferred to the therapist, who is unconsciously seen as the parent who promised fulfillment. The patient is prepared to do whatever the therapist asks under the illusion that winning the therapist's love results in self-realization. The therapeutic situation can become highly charged with these unspoken anticipations and, like the original infantile and childhood situations, it will end in failure and in the patient's feeling of despair about the therapy. Therapy is not a search for love but for self-discovery, or, one might say, for self-love. To whatever degree one is seeking to be fulfilled through a loving relationship one will be disappointed. Inevitably, the person will fall back into his despair. This constantly happens in therapy since only a person in despair would think that love and salvation lie outside of the self. If the patient can accept this fact, that his despair stems from an inner emptiness, the way is open to work through the despair to the fullness of being. In the succeeding chapters, we will look at this "way" to see what more is required to gain one's self.

CHAPTER 5

ANGER:
THE HEALING
EMOTION

In the preceding chapter I discussed the emotion of sadness, with special reference to its expression in crying. We have seen that all patients need to cry to discharge the pain and sadness caused by the physical and emotional "hurts" of their childhood. Children are taught not to cry and, in many cases, are punished or yelled at for crying. The inhibition against crying results in severe chronic tension in the muscles of the body's inner tube, which is concerned with the respiratory and alimentary functions. These tensions contract the respiratory tube, greatly limiting the person's breathing, reducing his energy and decreasing his vocal self-expression. But this is not the only effect of these childhood traumas. Severe tensions also develop in the muscles of the outer tube which has, as one of its main functions, to move the organism in space. The bodies of all patients reflect their painful history in the loss of gracefulness, in splits which separate the body's major segments—the head from the trunk or the pelvis from the thorax. These splits destroy the integrity of the personality, which cannot be restored simply by crying. The restorative or protective emotion is anger. All patients have considerable suppressed anger, in many cases amounting to a destructive rage, which they could not express as children when they were hurt. These feelings have to be expressed in a safe place if the body is to recover its vitality and its integrity. And, yet, as with crying, all patients have great difficulty expressing anger effectively and appropriately. Without this ability the individual is a victim or a victimizer.

Anger is an important emotion in the life of all creatures, since it serves to maintain and protect the physical and psychological integrity of the organism. Without anger one is helpless against the assaults to which life subjects us. The infant young of most developed species lack the motor coordination necessary for the expression of anger, and need the protection of parents. This is especially true of the human infant, who needs a longer time than most other mammalian infants to gain this ability. But to say that an infant cannot get angry is not quite true. Restrain an infant and you will sense his struggle to free himself, which represents an angry, though unconscious, response. Withdraw your nipple from a nursing baby and you will feel his gums bite down to hold the nipple if he is not ready to let it go. Biting is clearly an expression of anger as most parents know. As the child grows older and as motor coordination increases, his ability to express anger becomes more developed. He will respond with anger to any violation of his integrity or space, including personal possessions. If his anger fails to protect his integrity, the child will cry, feeling helpless against the trauma. The emotion of anger is part of the larger function of aggression, which literally means "to move toward." Aggression is the opposite of regression, which means "to move back." In psychology, it is the opposite of passivity, which denotes an attitude of standing immobile or waiting. We can move toward another person in love or in anger. Both actions are aggressive and both are positive for the individual.

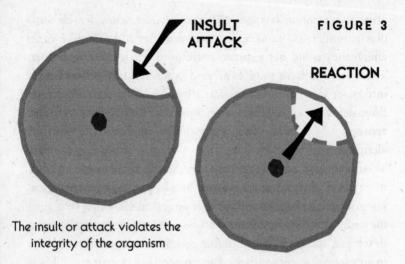

INSULT
ATTACK

FIGURE 3

REACTION

The insult or attack violates the
integrity of the organism

The reaction in the form of anger seeks to restore
the integrity of the organism

Generally we do not become angry at people who mean nothing to us or who have not hurt us. If they are simply negative we avoid them. When we are angry with people we care about, it is to restore a positive relationship with them. I believe we have all experienced the fact that after a fight with a loved one, good feelings are generally restored.

In a seminar at Reich's home in 1945 he stated that the neurotic personality only develops when a child's ability to express anger at an insult to his personality is blocked. He pointed out that when the act of reaching out for pleasure is frustrated, a withdrawal of the impulse takes place, creating a loss of integrity in the body. That integrity can be restored only through the mobilization of aggressive energy and its expression as anger. This would reestablish the organism's natural boundaries and its ability to reach out again. (See Figure 3.)

To the human being anger represents a surge of excitation upward along the back of the body and into the arms, which are now energized to strike. The excitation also flows over the top of the head and into the upper canine teeth, which are then energized to bite. We are carnivores and biting is a natural aggressive impulse for us. I have actually felt this flow of excitation into my canine teeth in an exercise with anger. As this excitation flows through the muscles of the back, they hump up in preparation for the attack. At the same time one can feel the hair erect along the head and back. We rarely see this in humans, but it is a common sight in dogs. This flow of excitation in anger is shown in Figure 4A. In Figure 4B, the flow of excitation is reversed, resulting in a widening of the eyes, raised brows, head pulled backward and shoulders raised. This is the energetic movement in fear. If an individual is unable to get angry, he becomes locked in a position of fear. The two emotions are antithetical; when one is angry, one is not frightened, and vice versa. By the same token, when a person is very frightened, one can assume that he has an equal amount of potential anger—of suppressed anger—in his personality. Expressing the anger releases the fear, just as crying releases the sadness. In most cases the fear is equally denied and suppressed, with the result that the person is immobilized or "dead." In this situation it becomes important to find a way to help a person get in touch with his suppressed anger.

Talking with a patient about his problems will on occasion enable him to get in touch with a feeling of anger, which he can express through a hitting exercise. A more direct way is through crying. If a patient begins to cry by using the exercises described in the preceding chapter he will feel his hurt and his pain. Often the sadness will change into a feeling of anger, which can then be expressed by hitting the bed. Just as one doesn't release all his sadness by crying one time, so no patient releases all his suppressed anger by hitting the bed one time. In the course of therapy, as the crying deepens the anger becomes stronger, more focused, better understood. It is also possible to mobilize the

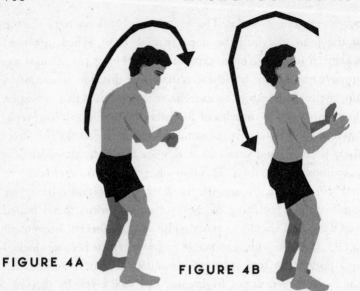

FIGURE 4A

FIGURE 4B

Direction of flow of the excitation
in the feeling of anger

Direction of flow of the excitation
in the feeling of fear

feeling of anger by doing the hitting exercise mechanically at first. Such an approach is like priming a pump: The action itself can induce a feeling of anger since the feeling is in the movement itself. In the hitting exercise, the person uses his fists if he is a man or a tennis racquet if she is a woman. The tennis racquet gives a woman a greater sense of power. Men have greater strength in their arms and can break a racquet hitting the bed with it. The patient is instructed to accompany the action with words which would also express his feeling. He could say, for example, "I am so angry" or "I could smash you" or "I could kill you." Combining words with the physical action focuses the feeling. Just as all patients have something to cry about or kick about in terms of their treatment as children, they also have much to be angry about. But their anger can also stem from a present-day situation which they were not able to deal with ap-

propriately because of their fear of retaliation. Since the exercise
frees up the tense muscles that have blocked the expression of
anger, it facilitates and promotes the ability to express anger in
all life situations. In my experience it never leads to "acting out,"
that is, expressing anger irrationally. And in all the years that I
have used this exercise with my patients, no one has ever been
hurt and nothing was ever broken in my office. If I sense that a
patient is losing control, I stop them and show them how to keep
command of their actions and still express their anger.

When I say that anger is not a destructive emotion, I am
making a distinction between anger, rage and fury. Rage is a
destructive action. It is intended to hurt, actually to break someone or something. It is also blind, and the attack is often against
an innocent, helpless person or a child. We speak of a person as
being "in a blind rage" or being "blind with rage." Rage is also
explosive, which means that it cannot be controlled once it blows.
One can contain anger but not rage. As I pointed out in my
book, *Narcissism*, rage develops when a person feels that his
power is thwarted or frustrated.[1] A child who persistently resists
a parent's demand can throw that parent into a rage aimed at
breaking the child's resistance, forcing him to submit. The failure
of a child to do what a parent commands confronts the parent
with his own feeling of impotence, stemming from the fact that
he was forced to submit when he was a child and was unable to
express his own anger out of fear. That suppressed anger now
becomes rage and is acted out on a child or other person of whom
he is not afraid. Many of my patients were forced to submit to
the power of their parents when they were small and often were
punished by spanking—a form of punishment particularly humiliating because it undercuts the child's sense of dignity and
privacy. Others have related how they were made to fetch the

[1] Lowen, A., *Narcissism, The Denial of the True Self.*

means of their punishment—a strap, a birch branch, etc.—which increases the fear and further humiliates the child. If the child is badly abused, the anger he would normally feel is buried under a mountain of fear and becomes destructive rage when it is released. Yet it has to be released before the person can feel and express genuine anger.

When I ask my patients to hit the bed with their fists or a racquet, what often comes out is rage, not anger. At first they are generally reluctant to put any feeling into their blows, which have an impotent quality. But once they begin to let go, they hit wildly and rapidly as if they want to smash or kill. This is a hysterical type of action in that it is not integrated with the ego and does not feel effective. When I ask what they are angry about, or against whom it is directed, they frequently say they don't know. These blows have little value, therefore, in further-ing the therapeutic process of self-discovery, but they are neces-sary to discharge some of the pent-up fury. Such actions are cathartic and constitute a safety valve in that they "let off steam." As the therapy progresses, both analytically and physically, the patient will get in touch with the reasons for his rage, his hitting will become more focused, and he will feel his anger. Saying the proper words while hitting makes the action *ego syntonic*. The expression ego syntonic denotes that the feeling and the action are in accord with and further the individual's sense of self. Too often a strong emotional reaction is seen by the individual as a loss of self and a loss of self control. Every patient that I have worked with has been hurt and humiliated to the point where the words "I could kill you" make sense. At the same time the patient is fully aware that it is a feeling he will not act out. The expression merely denotes the intensity of the feeling of anger.

A more intense anger than rage is fury. "I am furious" ex-presses an extreme feeling of anger symbolized by the whirlwind or tornado, which destroys everything in its path. One of my patients had a dream in which she felt a wind rising within her and lifted her from the ground. She also felt her cheeks puffed out with the wind as in the pictures one sees of the north wind blowing hard. While she floated off the ground, she was waving her hands, threatening some people in the room with her. I interpreted the dream as a rising wind which never broke loose, never became a whirlwind. This patient, whom I will call Susan, was terrified of her murderous rage. She had hit the bed many times in anger but it had never felt satisfying. Once while hitting the bed with the words, "I could kill you" addressed to her father, she froze temporarily in a catatonic stance, unable to move a muscle. Some years earlier another patient had related that she had once experienced a catatonic reaction when she had approached her brother with a knife intending to kill him. She said that something stopped her and she withdrew into another room where she stood immobile for almost half an hour in a catatonic condition. I realized that the catatonic reaction was the ultimate defense against the acting-out of her murderous impulse. Susan had told me many times that she was full of hate and often felt bitterly angry but could never express it. Her body had a frozen quality which she experienced as numbness.

That frozen quality is the physical aspect of hate. We hate deeply only those whom we once loved deeply but who, we feel, have betrayed us. But hate can be projected on others (transferred) with whom one has had no intimate or other personal relationship. Susan's relationship with her father was a mixture of love and hate. In the course of therapy she became aware that he had been sexually involved with her since she was a child. While she had no memory of any act of sexual abuse, she knew that he had looked at her as a sexual object since she was little. Even in his adult life, he regularly tried to press his body against hers when she visited her family. She recognized that he was seductive, that he was obsessed with sexuality, and at the same time that he derogated any girl or woman who manifested any sexual feeling. As a result of his behavior and her Catholic education Susan was ashamed of her body and terribly embarrassed by any sexual expression. She could not allow any sexual feeling to develop, and certainly not to show, with anyone. She was depressed and unable to mobilize herself for any pleasurable activity. On weekends she would spend most of her time in bed. It was only after several years of therapy that she expressed the thought that she could not go on much longer in her present state and that she could kill herself. Such an action represents the turning of her murderous hate against herself.

A frozen state can only be changed through heat—specifically, the heat of anger. Rage, as opposed to anger, is cold. An individual can sense the heat of his anger rise in his head as the excitation moves upward. He will become hot-headed due to the increase of blood in his head, which can also make him literally "see red." Anger is a life-positive force that has strong healing properties. On one occasion when I experienced such anger, it cleared up a sciatic condition that had troubled me for months. I have seen this also happen to one of my patients. In a similar way Susan's dream had a positive effect on her. Although there was no overt sexual abuse, she was tortured by the continuous psychological abuse of her femininity, which she had survived by numbing herself and cutting off all feeling. Any strong feeling could have overwhelmed her vulnerable ego. When she related the dream about a wind rising in her, she remarked that she thought it was a breakthrough. For the first time she was carried away by her anger and while it lifted her off the ground, she was not terrified. In the session following this dream she was able to tell me how much she appreciated my patience and support during the years when therapy made so little progress. And she also was able to tell me how much warm feeling she had for me. She had been too cold and too numb to allow such feelings to develop, and too frightened and too vulnerable to express them.

It should be emphasized that therapy aims to restore an individual's ability to feel and express anger, which is a natural response to situations that injure or threaten an individual's integrity or freedom. All children have this natural capacity to protect their integrity and freedom. Unfortunately, modern living conditions often force parents to frustrate a child's spontaneous impulses, which provokes the child to anger. He may strike at the parent, but despite the fact that such blows are not harmful, there are not many parents who will accept or tolerate such behavior. Most parents forcibly restrain an angry child, and many will punish him for what they regard as inappropriate behavior. Having power because of the child's dependence upon them, they can force the child to suppress its anger. This is most unfortunate since the child who is afraid to express anger at its parents becomes a crippled adult. Suppressed anger doesn't disappear. Children will act out the forbidden impulse against smaller children, deliberately hurting them. Or when the child whose anger is suppressed becomes an adult, he will act out against his own children, who are helpless.

Punishing a child for his expression of anger may be thought of as a way of teaching a child social behavior, but its effect is to break the child's spirit and make him submissive to authority. A child does need to learn the codes of social behavior, but this must be done in such a way that the child's personality is not damaged. In Japan I saw a three-year-old child hitting his mother, who made no effort to stop the child or reprimand him. In traditional Japanese culture a child is not controlled until the age of six, since his behavior up to that age is accepted as natural and innocent. After the age of six the process of socialization is done through shaming rather than physical punishment. In the upbringing of Spartan children who were trained to be fearless fighters, the child was not exposed to frightening situations or punishment before the age of six, to protect his spirit. Children whose ability to express anger is not undermined do not become angry adults. Despite the fact that they have a temper, they tend to be gentle people until they are abused or violated. Their anger is generally appropriate to the situation since it is not fueled by unresolved conflicts and past injuries. People who are hot-tempered or fly off the handle are sitting on a large amount of suppressed anger that is close to the surface and, therefore, easily provoked. Anger released through provocation does little to resolve the underlying conflict, which is the fear of expressing one's murderous feelings against the parent or authority figure who injured the child's integrity. That conflict is held and locked in the tension of the upper back and shoulders and can only be resolved when the anger is directed against the person responsible for the trauma. However, it is not acted out against that person since the injury is an old one. The appropriate place for such release is the therapeutic situation.

Many children are brought up with the idea that anger is morally wrong. One should be understanding, see the other person's point, turn the other cheek, be forgiving, and so on. There is much to be said for this philosophy provided one is not crippled or handicapped as a result. In most cases, however, the attitude of seeing the other's position amounts to self-negation, which stems from fear. It is a sign of graciousness to be forgiving, but the choice must be a real one. The individual who cannot get angry is not acting through choice but out of fear. It's been my experience that all patients have an inability to freely and fully express their anger. William was raised in a religious home where, he said, no one ever got angry. He claimed that he had never been angry in his life. His mother had raised him to be the perfect child, angelic and sweet. While he looked angelic at times, with his curly blond hair, he was not sweet. There was an unexpressed bitterness in his personality. He often complained of being frustrated in his career and in his love life. He was frustrated because he couldn't achieve the goal of being the outstanding person his mother had wanted and then when he accepted the failure of that ambition he was still frustrated because he wasn't free from his mother whose angelic little boy he still was.

William had never felt any joy in his life. Saddled with an impossible dream, he had been deprived of the innocence and freedom which is normal to childhood. It never occurred to him that he had a right to be angry at this deprivation. The result was that he struggled to find some joy in his work and in his sexual life, but this was impossible since struggling and joy are incompatible.

William needed his anger, for without the ability to get angry, he remained a victim, too vulnerable and helpless to give up his struggle, accept his common humanity and dethrone his mother from her superior position. He needed to feel how angry he was with his mother, but that would have been sacrilegious. Many patients report that they would feel guilty about expressing anger toward a parent, particularly a mother. Too many mothers inculcate a sense of guilt in their children for any negative feelings toward them. But guilt is based on fear and the suppression of anger.[2] Allowed to freely express his feelings, a child would retain his feeling of innocence. William had been a submissive child and had never expressed any negative feeling towards his mother. He had become her little angel after having been psychologically castrated and rendered impotent by his dominating mother, who also saw herself as a goddess in some way. It took William several

[2] See Lowen, Alexander, *Pleasure, A Creative Approach to Life* (New York: Penguin, 1975; Arkana, 1994) for an analysis of guilt, explaining its origin and persistence.

years of therapy before he was able to feel any anger toward his mother for the damage she inflicted on him despite the fact that he recognized he was hurt. While William's body was not frozen like Susan's, it was bound by so many tensions that there was little spontaneous movement and, therefore, little feeling of any kind. William operated largely through his will. Getting some vibrations into his legs was the first step in freeing him from the web of tension that imprisoned his spirit. It was a long time before William could cry. Fortunately, he kept at the exercises, in the sessions and at home, because they made him feel more alive. It was this commitment to his body that finally allowed him to feel some anger toward his mother.

One of the exercises that I encourage my patients to do at home is hitting the bed. I have done this exercise myself over the years to free up the tension in my shoulders and develop a freely flowing arm movement, which I believe is essential to the easy expression of anger. I was aware, at the beginning, that while I sensed power in my right arm, my left seemed weak and impotent. No person can be an effective fighter with only one arm. I used to hit 50 to 75 times every morning. Over time my left shoulder freed up and the blows with each arm became equal in strength and fluidity. But hitting the bed is not just a therapeutic exercise to free the arms from chronic tension. It also serves to release the tension that builds from the stresses of everyday life. We are not always in a position to express our anger at the time of the injury or insult. Sometimes one doesn't even feel the anger at the time of the insult because one is in shock. Some time later, as the shock wears off, one becomes conscious of how angry one really is about what happened. In some cases it is too late or impossible to express that anger to the person responsible for the injury, but one can vent the feeling of anger and release the tension by hitting a bed at home, and thus recover the integrity and good feeling one had lost.

Anger frequently flares in parents toward children who persist in doing what they want against the parent's injunction to stop. Children in our culture can drive parents mad when they cannot be controlled. In part this results from the overstimulation of children by the plethora of exciting objects in supermarkets and at home. In part it stems from the fact that parents are under considerable pressure to keep some order in their home and in their lives. They, too, are overstimulated and overwhelmed by their environment. The tension that builds up in the parent is often discharged by some physical punishment of the child. After venting the anger on the child the parent may feel sorry and guilty, but the damage is done. Reich had suggested that in such situations the parent go into a bedroom and let the anger out by beating the bed, not the child. I have recommended that action to all my patients. It relieves the parent and spares the child.

When they begin therapy with me, many patients report that they do not feel any anger during this pounding exercise. Every one of them had adequate reason to be angry at what was done to them as a child. Yet even when they recognize this fact, the anger does not flow because the tension suppressing the anger has not been sufficiently released. As a result, their movements are too segmented and too mechanical. An emotion is experienced only when the whole body is excited and engaged in the action. This means that the stretch of the arms over the head must be so full that it pulls the arms at their junction with the shoulders. I describe this to my patients as reaching for the thunderbolt. But for the whole body to be involved the stretch must actually come from the ground. For this to happen one bends the knees, lifts the heels slightly and stretches the body upward and backward from the balls of the feet. In effect, the body is arched like a bow anchored in the balls of the feet below and in the fists above.

When one achieves this position the blow is a free-flowing

movement. There is no more effort in hitting the bed than there is in releasing an arrow. Just as the power of an arrow depends on the amount of bend in the bow, so the power of the blow depends on the amount of stretch of the body. This is in accord with a physiological law which says that the power of a muscular contraction is directly proportionate to the degree of its stretch. Achieving such gracefulness in the action of hitting the bed is not easy for most people. The tension in the muscles of the shoulder, between the shoulders and the scapulae, and between the latter and the spine, are in many cases enormous and denote how severely blocked is the expression of anger. When one uses this exercise therapeutically, it is necessary to connect the tension to the psychological problem of guilt. This connection, however, can be more easily established after the person has experienced his anger. Joan, for example, became aware in the course of her therapy that men victimized her: they used her sexually. She could trace this to her relationship with her father who had been very seductive with her and at the same time showed her off to his male friends at the local pub. Connecting this awareness of her body, first through crying and breathing, then through grounding and hitting, gave her a sense of self that could support a strong feeling of anger. Hitting the bed strongly, she remarked that she could feel the warmth moving up her back. She added, "It feels good to have a back, and feel my backbone."

After this exercise she could understand why she had suppressed her anger. "When I got angry, my father became enraged and my mother blamed me. I learned to blame myself if I got irritated with or turned against someone. I wanted to be 'good'. Being good was my mother's idea of how to be. I was very religious as a young girl and being good gave my life meaning. If I was fresh with my mother I would feel guilty and confess my sin. It was my way of surviving but it left me crippled. Hitting the bed, I have a feeling of power."

In workshops where the group participates in all activities, it is possible to have the members release their rage rather quickly. In such workshops all feelings are heightened by the excitement that pervades the group when one member after another expresses a strong emotion. Thus, when one individual does the exercise of hitting the bed with anger, the others are motivated to follow. One after another each takes his turn at hitting and raging at his parents for the traumas suffered as a child. In almost all cases the rage is murderous, but it is quickly spent and the individual feels released. It is cathartic release. The person senses how angry he is but the anger is not discharged. Anger is not fully discharged until the tension in the muscles of the upper back and shoulders, which acts to suppress the anger, is released. But learning to hit the bed is an important step in that direction.

It has to be understood that the anger, while related to the past, stems directly from the existence of the chronic muscular tensions which bind the organism, reducing its freedom of motion. Anger is the natural reaction to the loss of freedom. This means that any chronic muscular tension in the body is associated with anger. Of course, if one doesn't feel the tension, one doesn't sense any anger. One accepts the limitations of movement and the loss of freedom as normal, much like a slave could accept the condition of his slavery without anger once he has accepted his loss of freedom. Once one feels and understands the tension, one becomes aware of how angry one really is and one realizes that hitting the bed to express anger is not a one-time exercise. It is done regularly both in the therapy sessions and at home, if possible, until arms and shoulders are free in their movements and the ability to express anger is fully restored.

Anger can be expressed by the voice in words or by the eyes with a look. But these modes of expressing anger are as difficult for most people as hitting. To let one's anger come through in a look requires that one feel the anger fully through the body, which allows the wave of excitation to reach the eyes. In some people, their eyes blaze when they are very angry. Cold, mean eyes are hostile, not angry, while dark, black eyes express hatred rather than anger. One can also use words to communicate that one is angry, but such words do not express anger unless said in an angry tone. That tone can be a quick, sharp sound, a loud shout or a scream. To express anger truly, the sound must be appropriate to the situation. Screaming and shouting, for example, often express rage and frustration rather than anger. It should be kept in mind that anger is not legitimately used to control others but to safeguard one's own integrity and good feeling. As adults we don't generally need to shout, scream or hit someone to express our anger. We can do it quietly, provided we feel it strongly. The previous exercise and others are designed to help patients feel their anger, gain the freedom to express it and then learn how to contain it and control it. Conscious control of feelings depends on the awareness of them.

I had been aware in my work with Reich that my ability to express anger was limited. I tended to avoid confrontation and to withdraw from a fight unless I was pushed to the wall. I sensed that there was considerable fear in me from which I could be free only by learning how to fight. The presence of this fear was responsible for my inability to maintain the feeling of joyfulness I experienced in my therapy with Reich. When I was a medical student at the University of Geneva, I made it a practice to hit the bed regularly each morning. I credit this exercise with greatly reducing the fear I would otherwise have felt in the face of studying and taking examinations in a foreign language. The exercise also had an overall positive effect on my health and mood and made my stay in Geneva an enjoyable one.

When I returned to the United States and began the development of Bioenergetic Analysis, I continued the practice of hitting the bed regularly in the mornings. In addition to the kind of hitting described above, where both arms are raised over the head and the blow is delivered with the two fists, I also began hitting with one fist after the other alternately, which is used in boxing and fighting. In this exercise I sensed that while my right arm felt strong and capable of delivering a good blow, my left arm felt weak and awkward in this movement. I could feel the tension in my left shoulder, which needed to be freed up. This happened gradually. I even set up a boxer's sandbag in the basement of my home so I could really punch it hard. But this exercise didn't do much for me. I wasn't trying to hurt anyone. I didn't feel angry. I was trying to free up my arms and regain my ability to fight. With that ability I would have no trouble expressing anger appropriately.

I learned later that the reason I didn't feel my anger at this time was that it was locked in my upper back, an area I was out of touch with. I became aware of this area when looking at some videotapes taken while I was teaching and working with patients. I saw that I was hunched forward and that my upper back was rounded out. It distressed me that I did not stand straight with my head up. I had on occasion described myself as an angry man, but I justified my anger by relating it to the senseless destruction of nature and the environment, which did make me angry. I was also angry at the blindness of people to the truth of their condition. But my anger had deeper roots, which I had been reluctant to face. I had been trying to prove to the world that I was right in how I saw things, that I was superior and should be recognized as such. But being right, feeling superior and achieving success did not lead to joy, only to continued struggle. And I was angry that I had been forced into this position to survive. This was not a healthy anger and I didn't need to hit anyone, to smash, to rage. I needed to accept my failure, to give up my ambition, to recognize and accept myself. I would then be free and not be angry. None of this happened overnight. Old patterns of behavior and ways of being change very slowly. But the slow change can have a dramatic aspect. One evening as I was being massaged, I explained to the masseur that I had considerable tension in my upper back related to the fact that I had a lot of anger. Then, without thinking, I said, "But I don't have to be angry anymore." As I said these words I felt my back literally "drop down" as if a weight had fallen off it. It was an amazing experience and I sense that since that day I have stood straighter.

Not being an angry person any longer, I find that I am much softer, more patient and easier. But, strange as it may seem, my ability to get angry, to fight, has increased greatly. Once expressed, the anger is gone. An angry person is a tense person which means that a tense person is angry. If the tension is chronic, the person is not aware of his anger. It can come out, however, in irritability at minor frustrations or in rage at major ones. It is not expressed appropriately in situations where it is needed. It can be turned against the self in self-destructive behavior, or it can be denied, leaving the person in a passive and submissive position.

Healthy children are quick to feel anger and to strike out when they are hurt or frustrated. As one gets older, one can contain the anger, when advisable, and not act on it immediately. And, as noted already, it can be expressed by a look or in words without the necessity for physical action. The ability to contain anger is the counterpart of the ability to express anger effectively. The conscious control necessary for containment is equivalent to the coordination and fluidity of the action expressing anger. Therefore, a person cannot develop the ability to control unless he also develops the ability to express. The exercise of hitting the bed can be adapted to serve both purposes.

Containment and control is developed as one learns to hold the excitation at a high level before discharging it, which is an adult ability. Children do not have the ego strength or muscular development to hold a strong energetic charge. When healthy children are hurt, their anger flares quickly and is immediately expressed. Adults should have the ability to hold their anger until an appropriate time and place is available for its expression. To contain the anger while doing the hitting exercise, one maintains the drawn-bow position for two to three breaths. The jaw is thrust forward to mobilize the aggressive feeling, and the eyes are open. In this position one inhales deeply through the mouth as elbows and arms are pulled backward for the blow. Instead of delivering the blow, however, one breathes out easily, releasing some of the tension in the arms and shoulders. With the second inhalation one stretches a little more and again releases by breathing out. As one breathes in for the third time, one makes a maximum stretch of the arms, holds the breath and the stretch for several moments and then easily lets the blow fall. No effort is needed for the blow since it is a release phenomenon. Trying to hit hard causes tension to develop and reduces the fluidity and effectiveness of the action. It is important to hold the elbows as close to the head as possible during the stretch, to engage and mobilize the muscles between the shoulders; with the elbows spread apart, the action becomes limited to the arms and does not release the tension in the upper back. Almost all patients need considerable practice to coordinate the movements and arrive at a free and easy swing in which the whole body is involved. When they do reach this point, they find that the exercise of hitting is pleasurable and satisfying.

This exercise is, in my opinion, the most effective means of reducing the muscular tension in the shoulders and upper back about which so many people complain. I have used it successfully to treat the problem of numbness and tingling in the arm and hand due to pinching of the nerve to the arm. This nerve passes through a triangle at the base of the neck as it enters the arm.

Tension in the muscles which comprise this triangle—specifically in the anterior scalene—is responsible for the symptom which is often called the scalenus anticus syndrome. When doing this exercise, it is not necessary to feel anger. Just as prizefighters practice hitting as part of their training and enjoy the activity, we, too, can find pleasure in the use of our body to express our natural functions.

However, when the exercise is used therapeutically to restore a person's ability to feel and express anger, it should be accompanied by words of anger. The words objectify the feeling and help focus the action. Saying "I am so angry" when one is hitting the bed integrates the mind with the body's action. Here, too, the tone of the voice reflects and determines the quality of the experience. To hit strongly but speak weakly denotes a split in the personality. The use of the voice resonates the central tube of the body and greatly increases the energetic charge of the action. The Japanese have long been familiar with this phenomenon and use a strong sound to effect a forceful action. Thus, they are able to break a solid piece of wood with the blow of the hand if they utter a forceful "Ha" sound at the moment of impact. How forcefully one says "I am angry" determines how strongly one feels angry. It is not the loudness of the sound that has this effect but the vibrancy and intensity of the tone. "I am so angry" said quietly but with intensity has a greater feeling charge than a loud shout.

Another exercise I use in group situations is having participants direct their anger at me. In this exercise the group sits in a circle while I stand or crouch before each in turn. I ask the participant to hold out two fists, thrust his jaw forward, open his eyes very wide and, while shaking his fists at me, say, "I could kill you." This exercise aims to bring a look of anger into the eyes, which is very difficult for most people to do. If someone complains that they do not feel angry at me, I tell them I do not take it personally. I say it is like acting—and actors should have the capacity to put real feelings into what they do. With my encouragement and the support of the group, almost all participants can feel some real anger. No one has ever attacked me, but I keep out of striking range and the fact that they are seated is further protection. When I do this exercise even alone, I immediately feel the hair on my neck and head begin to stand up. My ears draw back, my mouth goes into a snarl and I sense how easily I could attack someone. When I drop the expression, the feeling fades immediately. This confirms for me that the feeling is identical with the activation of the appropriate musculature. The inability of some people to mobilize their muscles is responsible for the absence of feelings of anger. It is equally true that the inability to activate the muscles that would produce the sounds of crying makes it very difficult for them to feel their sadness.

The eyes play a most important role in the feeling of anger. I have found that people whose eyes are relatively lifeless—that is, dull and without any spark—have great difficulty in feeling anger. I had a patient who was in this condition. It was very difficult to arouse any strong feeling in him. He was very bright and very much in control of what he did or said. This quality made him successful in his profession, but left him depressed. He suffered from headaches and often felt exhausted. This stemmed from the enormous effort required to stay in such control. Once while he was lying on the bed, I placed two fingers of my right hand on his neck at the junction with his head, at a place which would be opposite the visual centers in his brain. My left hand was on his forehead supporting his head. As I exerted a strong pressure with my two fingers against the base of his skull, I asked him to open his eyes wide and to picture the face of his mother. As he did so his eyes blazed and a fury erupted in him. He wanted to kill her. When he came up from the exercise I was amazed at the transformation that had occurred in him. He looked 15 years younger and his face had an aliveness that I had not seen in him before. His tiredness was gone and he felt energized; he told me that when he looked at the image of his mother, he saw a look of hatred in her eyes that had triggered his anger. I was hopeful that we had achieved a significant breakthrough and that the transformation would last, but it didn't. When he came for the session the following week, he was back to his tired, controlled self. He had had a vision of what he could be if he could fully mobilize his feelings, but the fulfillment of that vision was to take a long time yet. He still could not cry freely.

Every contracted muscle, every frozen area of the body, holds impulses of anger that is fundamentally the aggression needed to restore the integrity and freedom of the body. The arms and hands are our major aggressive elements, and very early in life the child learns to use them to express his anger. But hitting is not the only means of that expression. One could scratch, and there are many children who do that. Females are more likely to express anger by scratching, which may be one reason why we traditionally identify them with cats. Often to help a patient mobilize the energy and the feeling in his eyes, I will have him look into my eyes as I bend over him while he is lying on the bed. I can change the expression in my eyes at will from a soft look to a hard angry one, from a mocking expression to one of coldness. Most patients react to these expressions appropriately. On more than one occasion, when I let my eyes take on a seductive and mocking look, or a very hostile one, a female patient has brought her hands up in front of her face like claws and said, "I'll scratch your eyes out." We should never minimize the power of a look to frighten.

A third way a child can express his anger is by biting. Some young children are biters, which in almost all cases evokes a sharp and severe rebuke from the parents. Hitting can be tolerated, though it is not acceptable, but biting is never tolerated. It evokes a very primitive fear in people. The biting child is seen as a wild animal which must be tamed. However, we must recognize that it is a very natural impulse and that the best way to get a child to control it is by education, not punishment. Some parents go so far as to bite the child to let him know how painful it is, but they do it also to frighten him into never doing it again. The fear of biting then becomes locked into the personality in the form of chronic jaw tension. We saw in Chapter 3 that this tension is also connected to the inhibition of crying. It is the most common form of chronic tension in people and is responsible for temporomandibular joint pain, the grinding of teeth and, in my opinion, tone deafness. When the tension in the jaw muscles is severe, it can affect visual as well as auditory acuity. Tension in the jaw denotes holding on. We set the jaw in an attitude of determination not to let go, not to give up, not to surrender. In some patients the jaw has a grim look as if the individual is holding on for dear life.

While some diminution of jaw tension can be achieved through relaxation techniques, encouraging a patient to bite is the most direct approach to the problem. For this purpose I have them bite on a towel. In some cases this can evoke considerable pain in the tight jaw muscles but the pain leaves as soon as one stops the action. The pain is not a negative sign; the patient is attempting to mobilize very spastic muscles, which is necessarily painful. But with practice at home in biting and in moving the lower jaw forward and backward and from side to side, the muscles soften and the pain disappears. The grinding of the teeth at night ceases and patients find that they can open their mouths more fully and freely than before.

Sometimes I engage in a tug of war with a patient. Each one of us bites down hard with our back teeth on the ends of a towel and, like two dogs, tries to pull it away from the other. There is no danger to the teeth in this exercise if the biting is done with the molars. One can generally feel the tension extend from the temporomandibular joint around the base of the head. This tension, which circles the head at the base of the skull and extends into the temporomandibular joint, is the major resistance to surrender. It is the major mechanism by which a person holds on to control. It prevents him from letting go of his head and, therefore, of ego-control. Such control, when conscious, is positive, but in almost all cases it is unconscious and represents a holding-out of fear. Unfortunately the fear, too, is unconscious, which makes the problem inaccessible to a verbal approach. One of the fears is that if the person loses his head in a fight, he might bite his opponent or, perhaps, kill him. I shall discuss the treatment of this fear in a later chapter.

It often takes some time in the therapeutic process before the patient gets in touch with or senses his problem with anger. People believe they can get angry because they are easily irritated or blow up in a rage. After a year in therapy David remarked, "I find that I do not have my anger available. I need to be strongly provoked or pushed up against a wall before I bring it out." These remarks followed his complaint of feeling tension between his shoulders and neck. Since he was an active young man, he was surprised. He said, "Chopping wood never touched this tension." When a person suffers from chronic muscular tension in some part of the body, they move in such a way so as not to feel the painfulness of the tension. As one gets in touch with his body through the bioenergetic exercises, these tension areas are brought into consciousness. David remarked, "This week I felt my jaw as having been pushed back. The muscles from my jaw to my neck and shoulders feel very tight." As a non sequitur, he then added, "Last night I had a dream about my leg being cut off. I guess it was a castration dream." This made him think of his father and he said, "My father never expressed anger. He advised me never to fight in a game." David's awareness of the block to the expression of anger had a physical basis. He felt it in his body. "I feel that my head and neck are screwed into my trunk. I want to pull them out. I need to blow my stack." David was lying on the bed as he made these observations. I had him thrust his jaw and make a loud sound, which opened his throat. He then began to cry deeply. When it ended, he said, "My eyes feel lighter. My body feels more elastic."

In the next session the focus shifted to David's mother. As he was lying over the stool crying with soft sounds, he said, "I feel the tension in my lower back. It is so tight, so compressed. I feel like my mother is buttoning me up pretty tight." This feeling may have some reference to the experience of being diapered but I didn't make the interpretation so as not to interrupt the flow of his thoughts. He spoke of his longing to be physically close to her, then remarked that she doesn't allow anyone to make a real contact with her. He described her as "socially gregarious but not connected." He added, "I am important to her in terms of my achievements. I have to be there for her."

During an exercise designed to help him get in touch with the lower part of his body, he observed, "The lower part of my body is frozen. The upper part of my body feels like a closed tulip bulb wanting to open but not ready. I feel like my mother would attack my genitals to make me into a girl. She wanted a girl. She did not let me become a man. She castrated me psychologically. She was seductive with me but did not let me get close to her. I felt physically tortured."

I have detailed some aspects of this case to show the connection between anger and sexuality. Feelings of anger cannot be opened if one's sexual aggression is blocked. To the degree that a man or woman is sexually passive they are equally passive in the expression of anger. Their aggressivity is reduced in all areas

of life. Very often when an individual is able to express his anger
at the opposite sex (I am referring to true anger and not rage,
contempt or vulgar demeaning), he finds that his sexual feelings
are stronger and more aggressive. While anger, expressed in hitting, biting or scratching, is a function of the upper half of the
body, it requires a strong base of self-assurance and security to
be effectively expressed. A person who feels that he "doesn't have
a leg to stand on" can hardly be expected to feel comfortable
with strong, angry feelings. Lower back tension, which circles
the body and cuts off sexual feelings in the pelvis, also cuts off
the flow of energy to the legs and feet.

Actually, the bioenergetic work with the legs begins very early
in the therapy. A patient's introduction to the breathing exercises
over the stool is followed immediately by what is called a grounding exercise, in which he bends forward to touch the ground
with his fingertips. This exercise was described in Chapter 2. It
is referred to here again because of its central importance in
keeping a patient connected to his reality, namely, the ground
on which he stands, his body and the situation he is in. Anger
is a very heady feeling; it can overwhelm some people whose
egos cannot integrate the strong charge. Schizophrenic patients
can split off if the feeling of anger floods them. Borderline patients can become very anxious. This can be avoided if attention
is constantly paid to the patient's grounding. Whenever I sense
that the emotional charge of the exercise becomes intense to the
point where the patient may have difficulty in keeping his contact
with reality, I stop the exercise and ground the patient. This
reduces the charge in the body in the same way that the ground
wire in an electrical circuit prevents a blowout. But I would
emphasize that in almost every session in which one works directly with the body, the grounding exercise is part of the procedure. It gets the person in touch with his legs by increasing

the feeling in them, which then provides greater security and support for all self-expressive exercises.

We also can use the legs to express angry feelings, as in kicking, but it is not an expression of anger that is generally used by adults. Little children will kick their parents or friends in anger but adults rarely do so. Kicking is widely employed in the martial arts of the Orient more as a defensive action than an aggressive one. For one thing, when a leg is raised off the ground, one is unable to change position. When an adult kicks another person it is more an expression of contempt than anger. By the kick he is treating him as an unwanted object that is in the way and that he is trying to remove.

Kicking, however, has another more important function—as a way to protest. I discussed this expressive action in Chapter 3. Since it is also an angry expression and is so basic to the bioenergetic work I do with my patients, I will expand upon its use here. The expression "to kick" about some situation means to protest it. We all have much to protest concerning what has been done to us, and it is important to express this protest. In bioenergetic therapy, kicking as a protest is done with the patient lying on a bed, legs outstretched and kicking the bed rhythmically with the calf of the leg, one after the other. Generally we ask a patient to voice his protest as well as to kick. The simplest form of protest is "why." This exercise will demonstrate rather vividly a patient's ability to express his feelings. Many new patients find this difficult to do, and some can do the exercise but with little feeling. The former group consists of individuals who are able to respond emotionally to situations only when they are provoked. The spontaneous expression of feeling is alien to them. In this situation they do not see a reason to make a protest. The second group is afraid to express a negative or aggressive feeling. The inability of these patients to do the exercise properly should be analyzed in terms of the person's history. It can be shown to the patient that this inability stems from a childhood in which the expression of protest was not allowed.

This kicking exercise of protest is basic to bioenergetic therapy. If an individual cannot protest against the infringement of his innate right of self-expression, he becomes a victim whose orientation is survival, not joy. Once a patient accepts that he has a right to protest, the next step is to develop an ability to make that protest strong and effective. Some patients can use their voice strongly, but the action of their legs is weak and ineffective. In others, the kicking is fairly adequate but the voice is weak and unconvincing. This difficulty of coordinating both voice and movement denotes a split in the personality between the ego and the body, between the functions of the upper half of the body and those of the lower. No simple exercise addresses this problem as well as kicking. It is used regularly in the course of therapy to help the patient develop the coordination of the two halves of the body and gain the freedom to express anger strongly.

The problem of opening the voice was analyzed and discussed in Chapter 3, where the focus was on the ability to cry. But it is equally important for a person to be able to scream. Through crying one can mobilize feelings in the pit of the belly—deep gut feelings. The sound of such crying has a low, deep resonance associated with "letting down" or surrender. On the other hand, screaming is a high-pitched, high-intensity sound that resonates strongly in the air chambers of the head. It is the opposite of surrender and belongs, therefore, in the realm of anger.

In screaming, one "blows his top." The energetic charge that flows upward and results in a scream floods and momentarily overwhelms the ego. It is in some respects the opposite of the same charge that flows downward as sexual excitement to end in orgasm. In both of these actions the body is released from ego control, and both represent, therefore, a surrender of the ego.

Young children have very little trouble screaming because their egos have not yet taken full control of their responses. Women can scream much more easily than men for the same reason, but many are afraid to let go of ego control. Screaming is like a safety valve, allowing the discharge of an excitation that cannot be handled rationally. It can be used in this way to reduce an intolerable stress. I have encouraged my patients to scream whenever they feel too much pressure mounting inside. Again, the best place to do this is in an automobile on a highway where, with the windows closed, one can scream one's head off without anyone hearing.

The objective of therapy, however, is not just to free the voice but to coordinate the freedom of vocal expression with an equal freedom of physical expression in movement. The protest exercise is ideal for this purpose. The patient is asked to kick steadily, rhythmically and forcefully while saying "why," with the sound sustained as long as possible. When the patient runs out of air, he continues the kicking while taking two or three breaths before saying "why" again. In this second expression of "why," the voice is raised both in pitch and intensity while the kicking is also done more strongly. Again, at the end of the breath, the kicking is continued while the patient recovers his breath. On the third repetition, the "why" sound is raised to a scream while the kicking is as fast and as strong as possible. One seeks to yield fully to the expression of the protest. If that is achieved, the release is complete and the resulting feeling is joy. However, it is not easy to achieve; most people are too frightened to surrender fully to the body. In other cases the ego is too quickly overwhelmed, and while the patient may reach a scream, it is a dissociated expression—like a hysterical reaction which leaves a patient more frightened. In this case a patient may temporarily withdraw, curl into himself and cry like a child, after which self-control is reestablished. Such an experience is not negative since it allows a patient to gain the realization that the regression and withdrawal are temporary and that he needs more work to strengthen the ego. Patients who have been sexually abused as children have a tendency to withdraw or leave the body when feelings become overwhelming. Doing this protest exercise regularly strengthens the ego by connecting it more strongly to the body, thus reducing his tendency to split off from his body.

If kicking and screaming are integrated, the patient will not split off from his body. But for free and effective kicking, the legs must be relatively free from chronic tension. This is not common, since most people do not have enough feeling in their legs and feet and are not well-grounded. They are pulled up energetically into their heads and use their legs mechanically. They walk *on* their legs and feet rather than walking *with* them. Their legs are either too thin or too heavy. Kicking is one of the best exercises to get more energy and feeling into the legs. As an exercise I ask each patient to kick regularly at home in the same manner as they do in my office. I ask them to execute 200 kicks rhythmically, counting each leg as a separate kick. The knees are kept straight but not stiffly so, and the kick is done with the calf, not the heel. The leg should be raised as high as possible before each blow. Since this is an exercise to open the pelvis, no vocal expression need accompany the kicking. Most people cannot do 200 kicks without stopping, and few have trouble doing 100. Their breathing is not adequate for this exercise, but with practice, breathing becomes deeper and freer while the movements are easier. Like running, this exercise promotes breathing and is, therefore, an aerobic exercise. But in contrast to running, it is non-weight-bearing and places no stress on the knees. Furthermore, it can be done at home. People who have done this exercise regularly have reported important changes in their legs and bodies. The heaviness of the thighs from which many women suffer diminishes and their legs become more shapely. Breathing is also greatly improved.

"Why" is not the only word one can use with the kicking exercise. Saying "no" in the same way one says the "why" is an excellent means of promoting self-expression. Many people have difficulty saying "no," which undermines their sense of self. Saying "no" creates a boundary that protects one's space and one's integrity. Another good self-expressive statement is the phrase, "Leave me alone." This expression refers to the feeling very many patients have that their parents did not give them the freedom to develop naturally. In many cases parents demanded submission to their will, and when this was not forthcoming, became hostile and abusive. The parents saw their children's resistance as defiance of their authority, and were determined to break their will. In other cases, the parents were too involved with their children, whom they regarded as extensions of themselves. Too often, as we shall see later, this involvement had sexual nuances. Patients who have gone through such experiences need to voice their protest forcibly. Such statements as "Leave me alone," and "What do you want from me?"—when asserted strongly—help restore to patients the feeling that they have the right to be free, to be themselves, to fulfill their own being and not that of their parents.

Without that right a person's capacity to love is seriously handicapped. Too often the love that patients claim they have for their parents is the result of guilt rather than pleasure and joy in their relationship with their parents. One can't feel joyful in a relationship in which one can't be true to one's self. When parents give that freedom to their children, they receive their true love in return. But only parents who find joy in their relationship with their children can give them the love that supports the child's growth toward fulfillment of his being.

Patients are advised not to act out negative feelings toward their parents. Such acting-out is neither appropriate nor helpful. The traumas which patients suffered as children are in the past and cannot be redressed by today's actions. The past cannot be changed. But therapy can free a person from the restrictions and limitations on his being which are the consequence of past traumas. While these limitations can be greatly reduced by releasing and expressing the impulses which are bound up in them, this should be done in a therapeutic setting and not directly with the parents or others. An individual who is crippled physically and psychologically by the forced suppression of his natural impulses becomes free and joyous as his body regains its freedom and grace. Such a person can love truly, and may actually feel some love for the parents who abused him but who also gave him life.

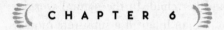

CHAPTER 6

LOVE:
THE FULFILLING
EMOTION

The Surrender to Love

Almost everyone has experienced the joy of being in love at some time in their life. Love has been described as the greatest and sweetest feeling, as the mystery which gives life its richest meaning. But when it is rejected or lost, it is also recognized as the source of our most intense pain. This is understandable, since love is a vital connection to a source of life and joy, whether that source is an individual, a community, nature, the universe or God. The disruption of the connection is therefore experienced as a threat to life. Since love is also an opening-up and expansion of the self to include the world, the loss of love results in a contraction and closing-off, which is as painful as the love was joyful. I have described the pain of this loss of love as heartbreak. Unfortunately, it can and often does last much longer than the joy of love because the individual becomes afraid to open up and reach out for love again. A longing for love persists in the heart but it cannot be fulfilled as long as the fear of loss or rejection persists.

The relationship which most symbolizes a loving connection

is that between a mother and her child. In the natural world a loss of this connection is fatal to an infant if a substitute mother cannot be found. When the relationship is secure the infant is fulfilled in its being and will develop into an adult who can establish a similar life-positive connection with another individual in the process of mating. The drive for a loving connection is so imperative that, despite the experience of heartbreak in childhood, an individual will seek, consciously or unconsciously, a loving connection with another individual. However, establishing the connection is not something one can consciously do. A person cannot will himself to love nor direct himself to fall in love. It happens spontaneously when individuals who meet suddenly find that their hearts are beating to the same rhythm and their bodies are vibrating on the same pitch. It can happen through eye contact or some other form of contact, but only when the charge in the connection is strong enough to set the heart beating, the pulse racing and the body vibrating with pleasurable excitement. It is the excitement of having found a lost paradise—the paradise that was lost when our loving connection to the mother was first broken.

No child can retain the loving connection to its mother indefinitely. Its destiny will force it to be separate, to move out into the world, to search for a mate with whom it will reestablish a loving connection that will be fulfilled in the sexual embrace and in the production of offspring. The child that is fulfilled on the oral level will be open to love and will move easily and securely onto the genital level.

The passage to adulthood takes one through a latency period during which the individual establishes positive connections to friends, then through adolescence, in which romantic love relationships are formed with the opposite sex. But, fulfilled or not, we all have to move into the adult position because of the biological imperatives of our nature. If we have been unfulfilled or deeply hurt in our childhood, our move toward a mature loving relationship will be tentative, our reaching-out hesitant and our opening to life constricted. We may fall in love because love is our lifeline, but the surrender proves to be only temporary, a momentary letting-go of ego-control in our continual struggle for survival.

This inability to surrender to love clear to the heart lies at the root of all the emotional problems people have and which they bring to therapy. The individual who has been hurt in his early relations with his parents has erected a set of defenses against being hurt again, the threat of which he perceives as life-threatening. These defenses are not just in his conscious mind, for if they were, he could surrender them at will. Having lived with them since childhood, they have become part of his personality, structured in the energy dynamics of his body. He has armored himself like a knight of old so that the arrow of love cannot pierce his heart. A more apt description is to say that he now lives in a closed world like a king in his castle, seemingly safe and secure as long as he retains his power, but cut off from the world of nature or natural feelings. He can venture forth into life but he does so as an excursion, with his guards within call. He has no faith in the love of his people, for bitter experience has taught him that betrayal is a constant danger. Like all human beings, he has a need for love but also believes that he has an equal or even greater need for power. For a king, falling in love is like falling off a horse: Should that happen, he will quickly remount to regain his position of power.

The analogy is valid since, in the hierarchy of personality functions, the ego sees itself as king. The king could say, "I am the servant of the people," but in reality his people serve him. The ego should serve the heart, but in most individuals love is in the service of the ego—to increase its power and sense of security. For many people love is a search for security as much as a search for pleasure and joy. As long as a person is needy, insecure or frightened, his reaching-out for love is contaminated by unfulfilled oral or infantile desires and is not a full-hearted sharing of pleasure and life. On the other hand, there are individuals who surrender their ego too quickly. These persons do not find the fulfillment that love promises because the surrender is to another person, not to the self. Without an ego, the person becomes a child who sees the other as a parent who can satisfy his needs, that is, who can provide fulfillment. One finds this kind of surrender in cults where, as I pointed out earlier, the member surrenders his ego and self to an all-powerful, all-knowing leader (obviously a substitute parent). While the surrender allows the person to feel free and joyful, it is based on a denial of the reality that the member is an adult and that the cult leader is an emotional child whose ego is overinflated with the illusion of omniscience and omnipotence. Inevitably the cult must collapse, leaving all parties devastated and disillusioned. This happens in marriages and love relationships, too, where the need to be fulfilled by the other is a dominant aspect of the attachment. Such relationships are described as dependent or codependent since each person needs the other. This does not mean that there is no love in such relationships, but the love has a childish or infantile quality.

The fear of surrender to love stems from the conflict between the ego and the heart. We love with our heart but we question, doubt and control with our ego. Our heart may say "Surrender," but our ego says "Be careful, don't let go; you will be abandoned and hurt." The heart as the organ of love is also the organ of fulfillment. The ego is the organ of survival, which is a legitimate function, but when the ego and survival dominate our behavior, true surrender becomes impossible. We long for the contact that would send our spirits soaring, make our hearts beat faster and set our feet dancing, but the longing is not fulfilled because our spirits are broken, our hearts are locked up and our feet are lifeless.

The excitement and warmth of love has a melting effect on the body. One can actually feel the sensation of melting in the pit of the belly when love is a major component of sexual desire. Love softens a person, but to be soft is to be vulnerable. People who cannot soften with love are said to be "hard-hearted," but the heart cannot be hard if it is to pump the blood through the body. The rigidity is in the voluntary musculature system that encases the body in an armor like that worn by knights of old. This rigidity prevents the person from crying deeply, from giving in to his sadness and, therefore, from surrendering to love. Because children can cry deeply, they can love fully. When we are cut off from the child we were, from the child in us, we are cut off from the ability to love. But this doesn't mean that we must act like a child. The surrender of the ego is the giving up of unconscious ego defenses which block the opening up and reaching out to life. However, I don't believe there is any individual who is completely incapable of feeling love.

In a previous book I reported the case of a young man who said that he didn't know what love was.[1] He was a narcissistic individual who functioned with very little feeling. He had cut off himself off from his body and operated solely from his ego. His body was so tight feelings and impulses were prevented from surfacing and reaching consciousness. But while it was very difficult for him to surrender to his body or to love, it was not impossible. As long as the heart beats love is not dead in a human being. The impulse to love may be deeply buried and strongly suppressed but it cannot be totally absent. This man came to me at the urging of a woman with whom he was sexually involved. She complained that he never expressed any feeling. He said that

[1] Lowen, A., *Narcissism, The Denial of the True Self.*

he didn't know what love was and asked me if love was what some people feel for their dogs. He claimed that he had not received any affection as a child, but this denial was a defensive maneuver to justify his closing up and to keep from feeling his hurt. He had buried his heart and his child, but both were alive in his unconscious. To free them from their living tomb would be a major undertaking.

The case described above is extreme. Most people feel some desire for love and can reach out for it in some measure. This allows them to feel some love, but since their desire is limited and their reaching tentative, they are not flooded with the excitement that would lift them to joy. They are too frightened to surrender fully, although in most cases, they are not in touch with their fear or their limitation. They are not aware of the tension in their body which restricts their ability to love. What they sense is a longing for love, which is not the same as the ability to love. When they meet someone who responds to this longing, they get hooked on this person like an addict or a cult member. They feel and believe that he or she holds the key to their fulfillment. And despite the pain or humiliation they may suffer in the relationship, it is very difficult for them to get free. This, I believe, is the normal pattern in our culture because the average love relationship is insecure and uncertain. And, since it does not fulfill the promise of joy which love offers, it breaks down eventually into disappointment and recrimination.

This longing for love represents the unloved and unfulfilled child buried within who, like Sleeping Beauty, is waiting for the prince to awaken it to life and love. The prince is the "good" parent with whom the child first experienced the joy of love subsequently lost. His lifelong search for that love is like the search for Shangri-la in James Hilton's novel, *Lost Horizon*. The seeker generally gets hung up on a person who resembles the "good" parent in some aspects but who also embodies many features of the "bad" parent who rejected or abused the child. Fulfillment cannot be achieved through regression, which can help one connect with the past and the child within. Once awakened and released, that child must become integrated into the adult life of the person.

For most people the issue is not whether they love or don't love but whether they can love with all their being. That would be too much to expect in a culture such as ours that regards surrender to the body as a sign of weakness. This half-hearted surrender to love defeats them, but, instead of recognizing the cause of their failure, they blame their love partner. True, that partner's commitment was equally half-hearted—for which he, too, will blame the other. Unfortunately, there is no way to make such relationships yield the joy each person looks for. A relationship flourishes only when both people bring a feeling of joy to it. Trying to find joy through someone else never works, in spite of all the love songs which hold out this dream. Love is a sharing, not a giving. A lover shares himself fully with the loved person. This will include both the sharing of joy and of sorrow. Since a pleasure shared is a pleasure doubled, the sharing of joy heightens that feeling, which can become ecstasy in the sexual embrace. The sharing of sorrow halves its pain. The joy one shares stems from the surrender to the body, not from the surrender to the other.

People do fall genuinely in love and experience the joy of surrender temporarily. It fails to hold up because it was more a needing than a loving, but this doesn't explain the fact that the person in love experiences it as genuine affection. My explanation is that falling in love has a regressive component that stems from the person's childhood, when such love was a total commitment. The person reexperiences the love he once felt for the mother or father, but in doing so regresses in part of his personality to being a child again. In this aspect of his personality he looks for the support and encouragement that he needed then. Thus, while the feeling of love is genuine, it does not stem from a surrender to the body and the self but from an abandonment of the adult position—standing on one's two feet, alone, and with responsibility for one's own good feelings.

This problem is well-illustrated in the case of Diane, an attractive, 40-year-old woman who was always ready to give herself in love and to take care of a man. In return she expected him to take care of her. It isn't that Diane was weak, helpless or incompetent. She had a strong and well-built body, she was intelligent and educated and she had supported herself in the past; but she was not fully connected to her body or herself.

When she first came to therapy, it was because she was married to a man who abused her physically and of whom she was afraid. Through the exercises described in the fourth chapter she developed a better sense of self to the point where she could stand up against her husband and eventually leave him. This first phase of her therapy ended on that note. She returned to therapy some four years later because she had gotten involved with another man who, while less physically abusive than her ex-husband, nonetheless treated her badly. In the interval between these relationships she had lived alone, held several jobs—none of which were secure or paid more than a subsistence wage—and had several affairs. Not long after she moved in with her new man, he opened a business and she went to work for him. He was older, had been previously married and had two grown children. Diane ran into some difficulty with the man's daughter, who didn't accept her—which was to be expected since she was seen as a rival. This situation was a repeat of her childhood, when she was seen as a rival by her mother. In neither case did she have the support of the man—her father or her lover. Once again she felt herself to be a failure, despite her feelings of love for the man and her sincere effort to help and work.

Something was wrong in Diane's personality, frustrating her deepest wish to find fulfillment and joy in love. She didn't complain about her fate, but she expressed her sadness that she didn't have any children. In therapy she did everything she could to become a more effective person. She did the breathing exercises over the bioenergetic stool, the grounding and the protesting by kicking, and she expressed anger at her man for the way he treated her. But, while these things made her feel better since she was more able to express herself, they did not produce a real change in her personality. She was stuck on trying to "make it" (therapy, job, love, life) and this hang-up was the very reason nothing worked for her. She needed just the opposite. She needed to accept her failure, to give up, to reach the place where she couldn't try, and she had to understand how and why she had become hung up on trying. One can't try to make life or love work. It is beyond trying. Diane needed to cry, to express her sadness at the failure of her life and her despair at ever being fulfilled in love. Her trying was a maneuver to deny her despair and the effect was to keep her hung up. She also had to understand why and how this dynamic developed in her personality. I have learned that one cannot depend upon a patient to reach that understanding by himself. A patient's characterological attitude has served several important functions in his life: It has been his means of survival, as I mentioned earlier; it has also served to give his life meaning and hope. These powerful forces fuel his determination to "make it work," to fulfill his hope. Since he will not accept that his hope is unrealistic and that the meaning he has assigned to his life is illusory, he will push on despite continued disappointment, which only seems to reinforce his determination. I believe it is the therapist's responsibility to confront a patient with the truth of his attitude. Of course, this is done sympathetically to help the patient gain understanding.

Diane's body told me that she was not a deprived child. She

was full-bodied and strong, which indicated that she had been properly cared for and nourished as an infant. Her problem stemmed from a later date—from about the ages of three to six—when she became aware of her sexuality and independence. In the case of a girl, those first sexual feelings are focused on her father. In the case of a boy they are focused on the mother. The girl loves her father and the boy loves his mother wholeheartedly, each with all the intensity of their youthful being. Though this love has sexual undertones, it is innocent, since knowledge of sexual intercourse is absent from the child's mind. Surrendering fully to the excitement of this love for the parent of the opposite sex, the child feels a joyfulness that provides meaning to its life.

Unfortunately, this innocent state of affairs doesn't last and the joy is lost. The parents become involved in the child's feelings, responding to their children with an adult awareness of sexuality which is not innocent. Generally, the parent of the opposite sex responds too positively while the parent of the same sex responds negatively. The father responds to his daughter's love not only as a father but as a man. His ego is flattered by her adoration and his body is excited by her warmth and vivacity. Since the same reaction does not occur with the mother, she becomes jealous and sees the little girl as her rival. This jealousy can be so fierce that the child is frightened for her existence. In self-defense she would like to destroy the mother, but she is helpless. Her father could be her protector, but dare he stand up to the mother's anger, being aware that he is emotionally involved in the triangle? His inability to protect his daughter leaves her feeling helpless and victimized. To survive she must cut off her sexual feelings, withdraw from the relationship with her father and submit to the mother. Diane did all of that.

The situation with a boy is not basically different. He is caught in the oedipal situation and set up as a rival to his father. If he surrenders fully to his love for his mother, he risks being taken over by her and becoming a momma's boy, which would alienate him from his father. To deny her is to risk her hostility and the withdrawal of her love and support, which he still needs. When a son is set up to be a rival to his father, he becomes vulnerable to the latter's jealousy and anger. He becomes afraid of his father, for he senses that to compete with him invites his hostility; not to compete is to lose his mother's love. Her sexual interest in him flatters his ego and excites his body and is very hard to resist. But to yield to the seduction and surrender to his excitement would lead to a sexual relation with his mother that is both too frightening and too dangerous. This happened to Oedipus who, in ignorance of his true identity, killed his father and married his mother. His fate was tragic. To avoid this danger, the boy must cut off his sexual feelings for his mother, which results in his being psychologically castrated.[2]

Diane's body showed the effects of the oedipal situation in her childhood. Although her body was strong and well-shaped, the lower part of her body, from her hips to her feet, was not strongly charged. While her legs vibrated when she was in the grounding position, the vibrations did not extend to her pelvis which was held very tight and rigid. The waves of her breathing did not extend deeply into her belly. There was no doubt in my mind that she was afraid to surrender to her sexuality. This fear was also manifested in tension in her chest, which restricted her breathing and limited her heart feelings. Her face had a youthful, almost childish, expression at times, which did not fit with her age. Diane was afraid to be fully a woman.

Her trying had a strong element of pleasing. She wanted to please me as she wanted to please both of the men she had been

[2] See Lowen, Alexander, *Fear of Life* (New York: Macmillan Publishing Co., 1981), for an in-depth analysis of the oedipal theme sociologically and psychologically.

involved with. They were both father figures for her, as they were at least fifteen years older than she was. This need to please preserved her role as "Daddy's little girl." Her hope was that in this role she would recover the love and the joy she had known as a little child with her father. But that same role prevented her from finding fulfillment as a woman.

Diane remembered the pleasure and joy she felt with her father. "He read to me every night before I went to sleep. He would read for a long time. It felt like an hour. I loved to listen to him. I couldn't wait for night when he would read to me. After reading, we would do windmills and I would go to sleep." He introduced her to literature and she remembers the walks they took together, when he would share his thoughts with her.

When I asked her about her relationship with the husband who was abusive to her, she said, "I loved to hear him talk. He was very brilliant." Sex with him, she said, was the best she had known. Since she loved him in spite of his physical abuse, I questioned her more deeply about her relationship with her father and she related a memory that never left her. "I remember lying in a bed when I was about three-and-a-half to four years old and feeling all open and joyful. These were not new feelings. On this occasion, I recall my father standing over my bed. I remember his hand hit me, but I felt he didn't want to do it. I don't know why he hit me. I was so happy to see him. It was such a big shock. It confused me and I've been confused ever since. I felt framed and now I'm always framed by people. I have to be on guard but I don't want to be on guard. I don't know how to protect myself."

The memory of being hit by her father, which she experienced as a betrayal, has disturbed her since she was small. She could never free herself from it because she could not express any anger against her father for the beating. In the session I suggested that she hit the bed with the tennis racquet to express some of this suppressed anger. At first I think she said she was angry just to please me, because when she tried to hit her father, she had trouble expressing anger. She admitted, then, that she had trouble expressing anger against him. She admitted also that she was afraid that if she got angry with him, he wouldn't love her. Since her father had been dead for several years, she was hanging on to the illusion that he still loved her.

She excused his attack by believing he was provoked by her mother. He was split between the two females. "His obsession with me made her extremely jealous and he had to choose." Then, for the first time, she recognized that his obsession with her was sexual: "He was too hot for me." But this did not diminish Diane's love for her father. His sexual interest inflamed her to a level of passion and joy which made her see this part of her childhood as idyllic.

The other side—her relationship with her mother—was hell. She had nightmares. When she was awake, her mother often beat her with a wooden spoon. Diane described her mother as a woman with an incredible iron will that was unbreakable. "She smashed my sister, who was beautiful and full of pleasure, as she danced about the house in high heels with her hair nicely done and her mouth made up. It was too sexual for my mother. She hit her with a wooden spoon and told her that she had to change her clothes or she would beat her to within an inch of her life. Now my sister weighs two hundred fifty pounds, is very affected in her speech and unreal." Diane was terrified of her mother— outwardly submissive but inwardly rebellious. She remarked, "I always felt like a Picasso painting, split right down the middle." Her protection, when she was around, was her Greek grandmother—her mother's mother—whom she saw as her best friend.

One can imagine the torment that Diane experienced as a child, torn as she was by her love for her father, with its sexual excitement, and her guilt and fear about that relationship. The guilt was overwhelming. "I felt responsible for what happened. If anything went wrong, it was my fault. It almost drove me crazy. I used to get terribly angry, but I couldn't get any results. I would hit my head against the wall and scream and scream. My anger became destructive. I wanted to break things, which made me feel more guilty."

In her late teens and after graduation from college Diane acted out her rebellion by becoming part of the counterculture, involved with drugs and sexually promiscuous. After a couple of years she realized that her behavior was self-destructive and went to Europe to study. There she fell in love with a nice young man close to her age who reciprocated her feelings. Unfortunately this relationship didn't go anywhere because his family objected to her background, which they saw as inferior to theirs. She had one other intense love affair with a young man which also failed to develop into a permanent relationship. Apropos of these relationships, Diane said, "I always pick sons whose mother's wouldn't let go of them. I had trouble with all their mothers. They were scared I'd take their sons from them."

In my view Diane is a pretty tragic figure, and she sees herself in somewhat the same way. She said, "I'm so unhappy. I don't see any future for me. I just get through one day after another." Those statements brought up some deep crying which then led her to remark, "There always is a deep sadness in me that I don't think will ever go away." Feelings don't change if one tries to overcome them. As part of her trying Diane put on a bright, happy face, partly to present herself as a positive, helpful person and partly to support her hope that she will find the love of her life. On a deep level it was a survival technique, for her sadness touched on a despair that she felt was life-threatening.

And yet this despair was no more realistic than the hope of

regaining the lost paradise that she knew and experienced as a child in her father's love. Both her hope and her love belong to her childhood and are irrelevant to her present situation. As a mature woman she wants a mature relationship with a man who would be more than just a lover, who would also be a companion and a husband who would work with her to build a home and perhaps raise a family. Since men need the same thing from a woman, it is not an unrealistic expectation. But it can only be realized if the woman and man are mature individuals.

Diane was not a mature woman. There was too much of the little girl in her, still looking for a father figure who would redress the situation of her childhood. He would adore her, tell her she was beautiful, affirm her innocence and protect her against the wicked stepmother. Of course, no man can do that for a woman. A lost innocence cannot be recovered. But guilt can be removed by restoring the fullness and freedom of self-expression, including the expression of sexual feeling. The fear of the stepmother can be eliminated by mobilizing the person's anger. That happened with Diane toward the end of her therapy, when she was able to stand up to her mother and ask for her help. To her surprise her mother proved very willing to help.

Diane's relationship with other men was more complicated, for she believed that in surrendering herself to them she was surrendering to love. That is how a little girl would see herself in relation to her father. He is her whole world and emotionally she exists largely in terms of that relationship. Hearing a little three-year-old girl scream with delight, "Daddy, daddy!" when she sees him, one can appreciate the totality of her surrender. Such behavior is characteristic of a child whose ego or sense of self has not yet fully developed. It is responsible for the deep sense of joy that a child knows, but we do not remain children. Between the ages of three and six the ego develops, and with it the sense of self increases, becoming a dominant aspect of the personality. In this period, known as the oedipal period, the child becomes aware of adult sexuality and loses his innocence. When at 6 he goes to school, joining other children the same age to learn about the larger world, he has or should have an established sense of self which we term the ego. He is now a fully self-conscious individual with a sense of pride in his individuality.

Self-consciousness is an alienating force in that it makes one conscious of being separate. In the home, one is part of the family and derives identity from his position in that group. That identity is relatively meaningless at school, where he is one of many children in the same position. In school he will form new bonds with one or more of his peers based on sharing a common situation and similar interests and feelings. These bonds can be very strong, and the feeling of love between two children can be very strong. The child still retains a strong attachment to family, but this love, if it is healthy, frees and supports the child in the move to establish a peer base. If the child is hung up on the family, as Diane was, he cannot move with ease into peer relationships. If deprived of love at home, he will become dependent and insecure with his new friends. If made to feel special at home, he will compete with and seek to dominate new acquaintances. In these two cases, the new friendships will not offer the joy he longs for.

The love of two children in a healthy peer relationship strengthens the individual's sense of self. It differs from the love of the child for the parent of the opposite sex in which, as we saw, there was a surrender of the self. Sex is not absent from these relationships since it is a fact of life, but its drive is greatly reduced so that the newly conscious sense of self can develop to mature proportions. Early in the development of analytic theory, Freud proposed the concept of two antithetical instincts which he described as the instinct for self-preservation, which he identified with the ego, and the sexual instincts, which can be described as the instinct for the preservation of the species.[3] That two such forces exist in the personality cannot be denied, however one describes them. In an adult they are polar forces representing an energetic charge in the body which pulsates between the upper and lower poles of the body, between the head with its ego functions and the pelvis with its sexual functions. Like any pendular activity it cannot be greater at one end than the other. Thus, in terms of energetic charge, the ego cannot be stronger than its counterpart, which is sexuality.

This principle may seem to be contradicted by narcissistic individuals in whom an exaggerated egotism is associated with decreased sexual potency. Such grandiosity, however, does not denote true ego strength; rather, the opposite. The overblown ego image is inflated to compensate for sexual impotence.[4] True ego strength is manifested in the regard of the eyes, which is direct, steady and strong. Such a regard stems from a high energetic charge in the eyes and corresponds to a similar charge in the pelvis. This equivalence is expressed in a folk saying, "Bright-eyed and bushy-tailed." Bright eyes also denote a strong ego, one that is grounded in the body and that stems from the feelings of pleasure and joy in the person. One can always tell when a person is in love by the fact that their eyes are shining.

Mature Love

Mature love is not a surrender of the self but a surrender to the self. The ego surrenders its hegemony of the personality to the heart, but in this surrender it is not annihilated. Rather it is strengthened because its roots in the body are nourished by the

---

[3] Freud, Sigmund, *Instincts and their Vicissitudes*, 1915, in *Collected Papers*, vol. 4, E. Jones, ed. (London: Hogarth Press, 1953) p. 67.

[4] Lowen, A., *Narcissism, The Denial of the True Self*.

joy that the body feels. In the statement, "I love you," the "I" becomes as strong as the feeling of love. It can be said that mature love is self-affirmative.

Diane is typical of so many of my patients who, in falling in love, surrender to the other person, not to the self. They give up their independence in the hope that the other person will take care of them. In effect, they regress to an infantile position which seems to promise the fulfillment of the relationship they had with the parent of the opposite sex. They become dependent and, in this position, are open and defenseless against abuse. Of course, such relationships very rarely last, and in the end the person suffers a repetition of the heartbreak experienced as a child.

It is impossible to have a mature love relationship unless one is a mature person, able to stand on his or her own feet, alone if necessary, and able to express feelings freely and fully. Such love is not selfish, for the person shares himself fully. It is self-centered, but this makes for an exciting relationship because each person is an individual with a unique self that they share with their partner. In such a relationship the fulfillment of love in sex is mutual in satisfaction and delight.

This view of love runs counter to the popular idea that in love one should be there for the other person. But this makes the relationship one of serving rather than of *sharing*. Sharing is between equals, but one serves a superior. Such love relationships lose their excitement soon and end with the partner who is being served looking outside of the relationship for the excitement of love that is missing in the marriage. When this happens, the spouse who is left behind tries harder to serve, to make it work, to be what the partner wants. Another of my patients, whose husband had left her, broke down, crying deeply, and said, "All my life I keep trying to fix things but it never works. You can't fix things. I keep trying to help people and all I get is hurt. I'm tired of it."

Philip was an attorney in his late fifties when he consulted me because something was missing from his life. He had been married at a young age and had three children with a woman he didn't love. But he stayed in the marriage for almost twenty years because he felt that his wife needed him. When I met him, he had been living for almost twelve years with Ruth, a woman much younger than he was. Their relationship had started as a sexual love affair, but over the past eight years both love and sex had faded away. They slept in the same bed, often naked, but there was no intimacy between them. Philip related that she often criticized him for minor lapses or weaknesses but, he said, he called her on it. Otherwise the arrangement was amiable and they seemed to manage well. They each had their own careers and were often apart on business.

It is easy to understand why Philip complained that there was something missing in his life. He had been in Freudian and Jungian analysis for many years in his search for fulfillment, which had also led him into meditation and other spiritual activities. For many years he had been part of a men's group aimed at raising the masculine consciousness. Philip was a man with a large, open face. He had a strong, well-built body and an attractive manner. Women were attracted to him but he was faithful to his partner.

To understand his problem one had to know his background. He described his mother as a domineering woman with hysterical tendencies and his father as a quiet, passive man. There were two children—his sister was two years older. He was aware that his mother had been seductive with him. She made him feel special and at the same time made him feel responsible for her happiness. With his background in analytic psychology, he knew about the oedipal issues and recognized that he was set up to compete with and outperform his father, which he did. He felt at ease in the world of men, where he could be aggressive with-

out being pushy. He had played college football. His problem
concerned his relations with women. But Philip's problem could
be worked through only if he got in touch with his feelings about
women, feelings he had strongly suppressed. He spoke very
openly and rationally about his problem, but with little emotion.
He came to see me because he realized that his emotions were
locked in physical tensions which needed a physical approach.

On the bioenergetic stool Philip's breathing was rather shallow. His chest was inflated and tightly held. Encouraging him
to use his voice helped somewhat, but it didn't evoke any sadness.
In the grounding exercise Philip had considerable difficulty getting his legs to vibrate. He did the kicking exercise with some
show of feeling, but didn't break through to an emotional release.
He had developed a strong control over his feelings early in life
and at this time it was beyond his conscious ability to let go.
Nevertheless, he felt much better after these exercises. He knew
they were taking him in the right direction and he was determined to stay with them and the analysis.

On one occasion when Philip was over the stool making a
continuous sound, his voice reached a point where it seemed that
it would break and he would cry. To my surprise he began to
laugh, which he could not stop. I have seen this happen to other
patients, and in almost all cases if the laughing continues, it ends
in sobs. It is an unconscious attempt to deny the sadness by turning it into a laughing matter. I joined in his laughter to help him
sense that his laughter was irrational, but all that happened was
that he laughed even harder until, after fifteen minutes, we just
stopped the charade. But while Philip did not cry that day, he
realized that he had a strong resistance to surrender and to letting
anyone "get to him."

Despite Philip's manly appearance, there was a boyish quality
about him that belied any claim to full maturity. Through the
analysis he became aware that he had felt trapped as a boy by

his mother and that he resented the responsibility she placed upon him to be her little man. Now, however, he was trapped by his narcissistic sense of being special and superior, which stemmed from his mother's sexual attraction to him. Narcissism is a common problem of men who had a seductive and controlling mother. There is a phallic quality in their personality related to their erective potency, which is the basis for their feeling of being sexually attractive to women. They see this erective potency as the ability to fulfill a woman sexually and emotionally. But for such a man, surrender to love is very difficult: on one hand it poses the risk of being possessed by a woman as he had been possessed by his mother, and on the other hand it would mean the loss of the phallic position, with its sense of specialness and superiority, since it would lead to a sexual orgasm that would discharge all the excitement of the seductive game. Philip told me that he could be erectively potent inside a woman for two hours while she experienced multiple orgasms. But Philip's failure or inability to surrender left him feeling unfulfilled and missing something important.

As I have said, surrender is not something one can do by an act of will, since it requires a giving up of the will. The will is a survival mechanism, and in Philip's case survival was seen as not letting any woman possess him. The turning point in Philip's therapy occurred shortly after the death of his ninety-two-year-old father, whom he had taken care of for several years. I had thought this event would have a freeing effect, since Philip's relationship with his father was mixed. He was the son, but in the later years of his father's life, he was also his father's father. His oedipal involvement with his father, which made him feel superior, also kept him the younger man. Now, he could claim the kingdom which is his full maturity. He became involved with a woman he had previously known, and their relationship became a passionate sexual affair, different in every way from the one

with Ruth. Philip now felt that he was truly in love with Elizabeth, the new woman. She was an older woman with grown
children. Where he had previously fallen in love with a younger
woman, Philip now gave his heart to an older one.

The circumstances of his relationship with Elizabeth made it
possible for him to lead a double life, spending the weekends
with his new lover and the rest of the time with Ruth. The new
relationship seemed to flourish, growing more intense over the
weeks, while the old one continued in its normal pattern. Philip
was aware that it was a situation that could not go on. Some
decision had to be made. His new love put pressure on him to
make their relationship known to his partner, but he hesitated,
unsure of himself. He described his conflict as follows: "I know
that she loves me deeply [speaking of the new relationship]. She
says that she has never experienced such sexual pleasure as she
has with me. We share so many interests and understand each
other deeply. I can be very open with her. She wants to be with
me all the time but I sense that there is some dependency in her
personality. With Ruth I have more freedom. Ruth is a practical
woman who knows how to get things done, which Elizabeth
isn't. But I love Elizabeth. I am sexually excited by her, which I
am not by Ruth."

Philip's personality wouldn't allow him to live a double life.
He had to be open and truthful with both women, but he knew
that telling Ruth about Elizabeth would hurt her and he could
not bring himself to do that. Philip was lying over the stool
breathing as we discussed this issue when he suddenly began to
cry. Some little crying had occurred earlier during the therapy
which, I believe, had helped open him to his new love. As he
was crying on this occasion, he said that he felt a pain in his
heart which he connected to the idea that hurting Ruth was also
his pain. His heart had been broken as he believed Ruth's would
be by his rejection of her. Philip began to cry more deeply as he

experienced a sadness that he had suppressed since childhood when his mother had rejected his sexual feeling for her. The tension in his chest that restricted his breathing and blocked his ability to surrender to love was his defense against the pain of his childhood and his vulnerability to being hurt again.

But the situation Philip was now in had no easy solution. He could not leave Ruth because he could not hurt her and because he was afraid to be alone. Ruth was in the same situation. Sensing that there was another woman in Philip's life, she could not leave. Knowing that their sexual love had dried up, she intimated that she was prepared to accept his need for a temporary liaison. She and Philip had not stayed together all those years out of love but out of need. Their relationship was one of codependency. Each needed the other. Caught as Philip was in his relationship with Ruth, he now began to sense that he was also being trapped by his relationship with Elizabeth. She put pressure on him to leave Ruth and threatened to end their relationship if he didn't, but she could no more give Philip up than he could leave Ruth. Philip became aware that Elizabeth was needy and that she would possess him as his mother had done. Philip began to realize that he would have to separate from Elizabeth for the same reason that he was separating from Ruth—namely, that he was not free.

To be free became the central issue in Philip's therapy now. Philip realized that he could not be a free person—that is, a person who is true to himself—as long as he was dependent. He was also dependent in his law practice, leaning on a partner whom he believed he needed. Thus, despite the fact that Philip was approaching sixty years of age, he was emotionally a boy and not fully a man able to stand alone on his own feet. Emotional maturity was the missing dimension in Philip's life, and that was a tragedy he could cry about and become angry about, which he did. In the course of the next year I could see a change in Philip's personality and in his life. He and Ruth separated, although they remained friends. He also separated from Elizabeth, although their sexual feelings for each other remained strong. And he assumed a leadership position in his firm. But what about love?

Philip said that he felt love for Ruth although he had no sexual desire for her. To put it simply, his heart was open to Ruth and in a different way it was also open to Elizabeth, for whom he still had sexual feelings. This was a love stemming from good feelings for these individuals and not out of any need of them. And his heart opened to include a sister from whom he had been estranged for years. And then, in a move that surprised me, during one session he said, "Dr. Lowen, I want to tell you how much I love you." He had a dream in which he saw himself ascending to heaven in a white cloud. He was tremendously excited, for he saw it as an expression of spiritual rebirth. At the same time he felt a deep, inner peace within himself which was also a joyful feeling. Despite the peacefulness and simplicity of these feelings, there was a passionate element in them. Philip was passionately in love with life and with the world. He had no need to search any further. He had found himself, he had reached into the core of his being, his heart, and there he had discovered the meaning of life in his openness to the excitement of being.

Philip had known love before. He had fallen in love with Ruth when they first became lovers, as he had with his wife when they first met. In those situations, the feeling of love was genuine but it didn't go deep enough and it didn't last. Just as one falls in love, one can fall out of love, and that happens too often because we are disappointed by the fact that the other person doesn't fulfill us. We do not realize that no one can fulfill us but ourselves and that our fulfillment stems from being fully open to ourselves and to life. When the arrow of love pierces our armor and reaches our heart, we are opened to life and joy but we do not stay open. Our egos slowly reassert their power, questioning, distrusting and controlling. The opening is seen as a breach in our defensive position which we must heal or close. Falling in love is not the answer, being in love is—that is, being open. First it is necessary to be open to one's self, to one's deepest feelings, and for that one needs to be free from fear, shame or guilt.

Fear undermines the ability to surrender to love. It is not a rational fear but one that stems from, and has meaning only in terms of, the individual's childhood experience. However, it still has power as long as we act as if we were in the same childhood situation. As long as Diane is trying to prove what a good little girl she is, being helpful, doing the right thing, she will be afraid to be herself, to accept her sexuality, to surrender to love. As long as Philip is afraid to be possessed by a woman, he will fight the surrender to love. He will base his appeal to a woman on his superior qualities, not on the fact that he is a man who needs a woman to fulfill his life. On this level Philip was still a young man who played at love and who still needed a mother to take care of him. He had never really lived alone. Since leaving his mother's house to get married, he had always been involved with some woman. Despite his love for Elizabeth, he knew that to move in with her immediately after leaving Ruth would be running away from facing his fear of being alone. As long as he felt dependent on a woman he would not be free; he would always fear her power to possess him. He would lack the maturity that makes the full surrender to love an expression of one's deepest self. Some weeks after the discussion of these issues Philip reported that he awoke feeling very joyful from a dream in which he felt that he was no longer frightened of being alone, no longer dependent on a woman. Every time a patient feels released from fear the experience is joyful!

Maturity is the stage in life when one knows and accepts one's self. One knows one's fears, weaknesses and maneuvers, and one accepts them. I don't believe we ever get to the point where we are completely free from the traumatic effects of the past, but we are not controlled by them. Acceptance doesn't mean helplessness. Since the problems are structured in the body in the form of chronic tensions, one can work with the body to free it up. The various bioenergetic exercises we use in the therapy can be done at home if a person knows how to use them.

Acceptance also means that one loses all shame about his difficulties or problems. Shame is similar to guilt in restricting a person's freedom to be himself and to express himself.[5] Diane's mother had made her terribly ashamed of her sexual feelings by labeling the innocent sexual behavior of a child common and dirty. But since these feelings were associated with very exciting and pleasurable sensations, the child was placed in a terrible conflict that almost drove her mad. She tried to suppress these feelings and, as we saw, did cut them off to some degree. But this built up an inner tension which then drove her to act the feelings out. All of us in civilized countries have some shame about the body and its animal functions, focused largely on sexuality, but few patients talk about their shame. They are too ashamed to talk about their shame and, being sophisticated, they deny it. Self-expression is not limited to feelings of sadness and anger. Most people have some dark secrets they are ashamed to reveal, and sometimes they even hide them from themselves. Fears, envy, disgust, repulsion and attraction, when hidden because of shame, become important barriers to the surrender to love.

Just as Diane suffered from shame, Philip suffered from a deep sense of guilt of which he was largely unaware. Guilt differs from shame in that it relates to feelings and actions which are

---

[5] I have analyzed the nature of shame and guilt in my book, *Pleasure* (op cit). They are called the judgmental emotions. Negative self-judgments are at the basis of both.

viewed as morally wrong rather than dirty or inferior. But most people who come to therapy today are psychologically sophisticated and deny any sense of guilt. Having denied it, one can't talk about it, making it difficult to free the person from his hang-ups. Children are made to believe that feelings of anger and sexuality are morally wrong when directed toward parents. Fear is associated with shame as well as guilt. Philip's guilt was manifested in the severe muscular tension in his body that held in a lot of sadness and anger but which only rarely reached consciousness. He had an enormous amount of anger against his mother for her betrayal of his love and against his father for leaving him in his mother's power. But he had bought into the deal and played the game, which made him feel special and superior. How could one get angry with a parent who treats one as special and superior? The anger will only arise when one senses the price one has paid in pain and frustration for this position. When Philip cried deeply, feeling the pain in his heart, he was on his way to becoming a free man.

The surrender to love involves the ability to share one's self fully with one's partner. Love is not a matter of giving but of being open. But that openness has to be first with one's self, then with another. It involves being in touch with one's deepest feelings and then being able to express those feelings appropriately. For Philip it meant the realization and acceptance of how angry he is with women—all women—for each one represents his mother in some way. For Diane it meant the acceptance of her anger toward all men, including her therapist, for each one represents the father who betrayed her. The surrender to the body and its feelings is the surrender to love.

CHAPTER 7

THE BETRAYAL
OF LOVE

As patients get more in touch with themselves and the events of their childhood, they generally become aware of feeling betrayed by their parents. The feeling of betrayal then gives rise to intense anger. After two and a half years of therapy, Monika said, "I feel so betrayed by my father. He used me. I loved him and he used me sexually. When I connect with my pelvis I feel how much I've been betrayed. I don't understand why men do this." Then she added, "I feel like an animal. I feel so angry. I want to bite but I am afraid to focus this feeling on the penis."

Her feelings were triggered by the recent breakup of a relationship with a man to whom she had given her love. He accepted her love but was often critical of her. To accept a woman's love without returning the love or showing respect is to use her. Her father had used her by being seductive with her, exciting her love and then showing her off as a sexual object to his male friends. Regardless of whether this behavior can constitute sexual abuse or not, it was a betrayal of the love and trust a child has for a parent. Of course, every act of sexual abuse by a parent or an older individual with a child is a betrayal of trust and love. But I also believe that every act of betrayal has within it an element of sexual abuse, whether overtly acted out or covertly suggested.

Another patient, a man, experienced this feeling of betrayal by his mother. As a child he could not stand up to her. She tried to control almost every facet of his life and behavior, with the result that, as an adult, he could not act in his own best interest. He had to be successful, to do the right thing in society's eyes, so that his mother would be proud of him. He had been her little "house boy," and as a man he functioned in a similar capacity with his wife. One session he came in complaining of a dry throat; he couldn't make a loud sound nor get a good breath. He felt choked, and the image that emerged was of a dog being led about by a collar and leash. In this case, it was a choke collar. She dressed him well and paraded him about as one might a prize poodle. Realizing this he said, "I had to make her feel proud, to fulfill her image of the superior mother."

She used him, just as Monika's father had used her, to gain some sexual excitement and the satisfaction of possessing an adoring child. She was completely unaware that in this behavior she was robbing her son of his manhood. Her actions represented the need to do to a male what had been done to her as a child. As we saw in the chapter on anger, one acts out on those who are helpless and dependent the insults and traumas one received when one was helpless and dependent as a child.

The use of power against another always has sexual undertones. Parents use their power to discipline a child so that he becomes a "good" child and later a "good" adult. To be bad, on the other hand, is not just to be negative or hostile, but to be sexual. A "good" child is submissive, doing what he is told. He is told that such behavior will gain him love, but it is a false promise, for all he will get is approval, not love. Love cannot be conditioned. Conditional love isn't true love. In defense of parents it must be recognized that some discipline must be enforced to keep some order in the home and to protect a young child from hurting himself. But discipline is one thing and breaking a child is another. People who come to therapy are individuals whose spirit has been damaged or broken. This is also true of many who do not come to therapy. Without thinking, most parents will treat their children as they were treated by their parents. In some cases they do this despite an inner voice that tells them it is wrong. An abused child often becomes an abusive parent because the dynamics of this behavior have become structured in his body. Children who have been subject to violence are generally violent with their own children because the latter are easy objects for the release of the suppressed anger. In time, children identify with their parents and justify such behavior as necessary and caring.

The following account of a session with one of my patients illustrates the perverse relationship that can exist between a parent and a child—in this case between a mother and daughter. Rachel was a woman in her early forties who consulted me because she was depressed. At the time she was in an ongoing therapy with an analyst in her own state. She had met me in a workshop and was intrigued with the idea of working with the body to resolve her problems.

Rachel was an attractive woman, above average in height, with a slender, well-shaped body that, however, did not seem strongly charged. Her face had a young look that denoted a childlike quality in her personality. Her legs were thin and not strong-looking. This was the third or fourth session I had had with her.

She began by saying that she had been through a very difficult three months since our last visit. "I was in a real severe depression and I was really scared that I was never going to be able to throw it off. I think I may be getting closer to that part of myself that is stubborn and resisting. When I thought about going to see you, I dreaded it. I look forward to seeing you but I dread the physical work. When I was here the last time I curled up on that couch, in a fetal position, and it was all I wanted to do."

She continued, "In my therapy I've been dealing with dreams, one of which was about snakes. I have a recurring snake image in my dreams. I dream about a lot of snakes, but this particular snake image recurs. The snake was hanging in a doorway, all coiled up and menacing. It was a big snake, like an anaconda or python, which would encircle me and squeeze me to death. In a recent dream I saw the snakes in a museum behind glass and I forced myself to look at them. Two others had primate-shaped skulls instead of snake-shaped skulls. They were becoming more human.

"At this time in my therapy, I was dealing with the pain of my brother's molestation of me. When I thought about the dream I had a vague sense that the museum was in Philadelphia, like the Philadelphia Museum of Art. As I wondered about Philadelphia, it occurred to me that it was called the City of Brotherly Love. I thought that brotherly love got all twisted with sexuality."

When I inquired about her brother's age, she said he was four years older and added, "I loved him so much I would have done anything he told me to do. When I was describing what he did to me to my other therapist, I felt I would pass out. I toted it around all these years, unable to put into words what he did. I was afraid people would dismiss it as a common occurrence and I would be terribly ashamed.

"I was so concerned about myself that I went for some psychological testing. I took the Rorschach test and saw female sex organs everywhere. There was a clearly phallic symbol at the top which I pointed to, saying, 'I don't know what this little gizmo would be.' The examiner chuckled."

When I asked Rachel what her analyst thought about her

seeing vaginas in the blot, she said that she had just received the results and they had not met since. I thought that Rachel's seeing the vaginas had something to do with her mother. I asked her to tell me about that relationship.

She said, "Well, I . . . I . . . I . . . have this feeling that my mother sexually abused me."

I had a similar intuition.

Rachel added, "I told my analyst about an incident that happened when I was a child. I had a thorn in my leg or something and I wouldn't let my mother get near me to touch me. She forcibly took me in her lap and I was screaming, 'Someone come and help me.' I was terrified. I have this sense of aversion to her and there is a definite sexual tone to it."

Rachel described her mother as being the powerful one in the family. "She ran it and she divided us against ourselves so that we would have no one to turn to. I . . . I am trembling inside telling you this."

I complimented Rachel on being courageous enough to face these issues, to which she replied, "I am, but I think it's almost to my detriment. A friend told me that he thinks I would walk into the mouth of a lion. I think I'd walk in with some kind of jack to catch the jaws."

The vagina is like a mouth, I pointed out. It swallows you. "And you sensed your mother's desire to possess you?" I asked Rachel.

"Yes, not only to possess me but to obliterate me."

"Did you sense her hostility to you? Did you think she could kill you?" I asked.

After a long pause, she answered, "Well, for one thing, she whipped me almost every day."

I was shocked. I commented, "That must have been done to make you submissive, to break your spirit."

"I have had fantasies where I wondered how far she would

go," Rachel said. "There would be a breaking point in the whipping. I would make a decision not to cry. I wouldn't give her the satisfaction. But then I'd cry just to make her quit. In a childlike way I was afraid that if I didn't she would kill me. I could sense in her a growing erosion of her control—a growing rage against me for not breaking."

At this point I had the strong feeling that Rachel's mother was sexually involved with her. I suggested that her mother's behavior had a lesbian aspect.

In a low, soft voice Rachel said, "I'm glad you put the word on it." Then she added, "I think she was jealous of me because she had a very hard childhood. I think she was sexually abused. She is a big woman with large bones. I came into the world a slender, feminine creature. I think it bothered her."

I pointed out to Rachel that her mother identified with her femininity and wanted to possess it. Rachel said that her mother was a very masculine woman, big and fat. Her mother, she said, used to make excuses to look at Rachel's vagina.

At this point, Rachel complained of feeling woozy and faint. She sighed and moaned, "Oh, God."

Rachel remarked that her mother gives her "the creeps." She said she was repelled by her mother and still is, that she could hardly bear to be around her. Then she related an incident which demonstrated the evil power her mother had over her. "When I went to Germany and had my baby, I was able to breastfeed it successfully. Then my mother came to see me, and the day she arrived my milk dried up and never came back. Boom, just overnight."

Then Rachel said that she believed that what her brother did was an acting out of her mother's feeling, not her father's. "My mother condones that. She gets a lascivious kick from it. I guess she's projecting her own self-hatred onto me for being sexual and giving me the message that I'm dirty and seductive. But I was not seductive. I went out of my way not to be seductive. I wanted to be clean and innocent, not to know about sex. I was not aware that what my brother did was sex. I just knew it was intrusive, frightening and dirty and I didn't like it."

After a short pause, Rachel remarked, "I feel such a sense of relief. I know it's true." We say that the truth can set one free. But that happens only when one accepts the truth. Acceptance involves surrender, a surrender to reality, to the body, to one's feelings. Rachel had never surrendered, she had never given up her struggle to escape her mother, to escape the past. This struggle had enabled her to survive but it also kept her tied to her childhood. And since it is impossible to escape the past, the effort to do so is doomed to failure, leaving one with the same feelings of hopelessness and despair one knew as a child. The idea that one can escape the past is an illusion which collapses again and again in the face of reality, leaving the individual depressed.

Rachel, like all survivors, keeps trying to change the past, to find the love that would save her and restore her self-esteem. It is the story of Sleeping Beauty, upon whom the wicked witch put a curse, condemning her to sleep for one hundred years—in effect, removing her from life—and surrounding her castle with an impenetrable thicket of thorns. What saved Sleeping Beauty was the love of a handsome young prince who had the courage to get through the thorny barrier and wake her. It is also the story of Cinderella, who was rescued from the life of a scullery maid by the love of a young prince. In the Cinderella story, a good witch provides the means whereby the beauty of Cinderella can be seen by the young prince. Both of these stories represent the dreams of young girls to be saved from the evil power of a wicked witch, or an evil stepmother. But every mother who turns against her daughter out of jealousy becomes a witch or an evil stepmother.

Like Diane, whose case I presented in the previous chapter,

Rachel was involved with a man who took care of her financially but abused her sexually. He was supposed to be the white knight, the good father who would love her and protect her from her cruel mother. But her dependence on him kept her in the role of the princess, the frightened little girl who sees her mother as all-powerful. Rachel was aware of this, for she said, "I am not ready to get out, start kicking and earn a living. And I hate myself for that."

Realistically, both Diane and Rachel are competent individuals who can and have supported themselves. It seems to me that there is something perverse when such a person stays in an abusive relationship. On one level it represents the acting-out of self-destructive feelings that stem from a deep sense of guilt and shame. Both Rachel and Diane believe that they are not worthy of a man's true love because they are not clean. They have been "dirtied" by their exposure to adult sexuality when they were still innocent. This deep guilt blocks their surrender to their own sexuality, which is the natural avenue for the expression of adult love. In place of the surrender to the self they surrender to a man, which allows them to sense some joy and to believe they do love. But these relationships do not work. They repeat the childhood experience with the father—the surrender and the betrayal. The repetition—compulsion, as Freud called it—has the force of fate.[1] It is now a well-known maxim: "What we do not remember, we are forced to repeat."

The woman is betrayed by the fact that the man she loves is not a knight in shining armor but an angry male who himself feels betrayed by women. His history would reveal that he was betrayed by his mother who, in the name of love, used him and abused him. Now he is being used by another woman who ex-

---

[1] Freud, Sigmund, *Beyond the Pleasure Principle* (New York: Liveright Publishing Corp., 1950).

pects him to be her savior, her protector and provider. At the same time he finds that he is sexually involved with a little girl, not a real woman. On some level he feels cheated and that triggers his anger, while, on another level, he senses the power to hurt her and to abuse her. He will, consciously or unconsciously, act out the hostility he had toward his mother on his partner, who will submit in the hope of proving that she is not like his mother and that she truly loves him.

The motives behind such self-destructive behavior are complex. Were Diane and Rachel simply being masochistic in allowing themselves to be abused? Masochistic behavior is itself very complex, for the true masochist claims that he derives pleasure from his abuse, which I believe. Wilhelm Reich did an analysis of this seeming anomaly.[2] In a case involving a male patient who could enjoy sex only after he was beaten on his buttocks, Reich showed that the beating removed the fear of castration, which allowed his patient to surrender to his sexual feelings. In the patient's mind it could be something like, "You beat me for being a bad boy but you won't castrate me." Because of the endemic nature of the oedipal problem, a fear of castration exists in almost all men in our culture. The fear of castration is associated with guilt about sexuality, but in only a few cases is the guilt so strong that it drives the individual into a masochistic position.

But while this analysis is valid, it does not account for the feelings of love which both Diane and Rachel expressed for the abusive men they were involved with. I have to believe that those feelings were genuine and that without them they could not submit to the abusive treatment. The idea that one can love his persecutor is not so strange when we realize that in childhood the persecutor was also the loving parent. Rachel's father loved

[2] Reich, Wilhelm, *Character Analysis* (New York: Orgone Institute Press, 1945).

her despite the fact that he was seductive with her and could not stand up for her against his wife. Diane's father was a source of joy to her when she was little and she loved him dearly. As a loving parent he promised to be there for her in her need. It was his failure to live up to that implied promise that constituted the betrayal. In the next chapter we shall see that this is true even of the father who sexually abuses his daughter.

A child is trapped by such a betrayal because he senses that the betrayal is more the result of weakness than it is an expression of hostility. With his deep sensitivity a child can sense the parent's love even when he is being hurt. The child senses the feelings that are below the surface and trusts them. It is as if the child believes that the abuse is an expression of love. Rachel believed that her mother loved her, albeit in a perverted way, and that the beatings were an expression of her sadistic love. "You wouldn't hurt me if you didn't care" is a strong conviction in children. A child might say, "If it is true that you love me, why can't we make it work? I'll do everything I can to help." In effect the statement says that the child is prepared to surrender himself to gain that needed love.

If we remember that a child is innocent, we can understand that he cannot comprehend nor deal with evil. Yet we would be naive not to recognize that evil exists in the human world. It does not exist in the natural world since these creatures have not eaten the fruit of the tree of knowledge and do not know good from evil. They only do what is natural to their kind. Man ate the forbidden fruit and is cursed with the existence of evil, against which he struggles. In some people the evil is so strong that it can be seen in their eyes. Many years ago while my wife and I were riding in a subway train, we happened to look at the eyes of a woman sitting opposite us. We were both struck by their evil look. Since we both saw it we could not doubt the truth of our impressions. I have seen that look only very rarely in other people, but one other case struck me forcibly. A mother and daughter consulted me about the girl's condition. My evaluation of the daughter supported a diagnosis of borderline schizophrenia. In the course of the interview during which both were present, the daughter made some negative remark to her mother. The latter looked at the girl with such a black look of hatred that I was shocked. It wasn't a look of anger or even of rage, but of pure hatred. If looks could kill, this look could do it. It was so destructive. But this mother professed love for her child, which was a denial of her true feeling. No child could deal with such contradictory messages and retain his sanity. This mother had an evil side to her personality which she covered up in words of love and caring. Her evil quality stemmed from the denial of her hatred.

Hatred is not evil any more than love is good. They are both natural emotions which are appropriate in certain situations. We love the truth, we hate hypocrisy. We love that which gives us pleasure, we hate that which causes us pain. There is a polar relationship between these two emotions just as there is between anger and fear.[3] We can't be angry and frightened at the same moment, although we can oscillate between these feelings as the situation requires. Thus, one moment we are angry and prepared to attack, then that impulse collapses and we feel frightened, wanting to withdraw. So we can be loving and hateful, but not at the same time. The anticipation of pleasure inspires us and draws us out. We expand and feel warm. If the excitement mounts, we feel loving and receptive. Should we be hurt in this condition, the body contracts and withdraws. If the hurt is severe, the contraction produces a cold, frozen feeling in the body. To produce such a strong contraction, the hurt must be inflicted by someone we love. Hatred can be understood, then, as frozen love.

[3] See Lowen, A., *Pleasure*, for a full analysis of these relationships.

In a session with a parent and child, I have heard the child scream at its parents, "I hate you, I hate you." Having expressed his hatred, the child bursts into tears and ran into the parent's arms. If hatred is frozen love, it explains the facility for one feeling to turn into the other. We can't hate if we can't love and vice versa.

When we are hurt by someone we love, our first reaction is to cry. As we have seen, this would be an infant's response to pain or to distress. An older child would more naturally react with anger, to remove the cause of the distress and regain a positive feeling in his body. The aim of both reactions is to restore his loving connection with the important people in life—parents, other caretakers and playmates. If that connection cannot be realized, the child remains in a state of contraction, unable to open up and reach out. His love is frozen; it has turned to hate. If the hate can be expressed, as the little girl did to her mother, the ice is broken and the flow of positive feeling is restored. However, just as few parents tolerate a child's anger, so even fewer would accept a child's expression of hatred. Unable to express the hate, the child feels bad and perceives himself as being bad—not evil, just not a good child. The parent who caused the child all this trouble is seen as good or right, to whom obedience and submission is due. This submission becomes a substitute love. The child will say, "I love my mother," but on a body level one can see the lack of any feeling of love—no warmth, no pleasurable excitement, no reaching out. It is love out of guilt, not out of joy. The child feels guilty for hating its mother.

In subsequent sessions, Rachel expressed her reluctance to see her mother, with whom she was still involved. She felt that her mother still had some power over her and that she was not free, more like a puppet than a person. But she could not mobilize any anger against her mother; she was too guilty and too frozen with fear to stand up to her. On some level she experienced her mother as a witch. Certainly, her behavior toward Rachel was inhuman. I am sure that she had some love for her child, but in her attacks upon the girl she seemed possessed by some evil spirit. At those times she hated her daughter and could have destroyed her. No doubt she had been similarly treated, and the hatred she felt toward her daughter was a projection of her hatred against those who had abused her. In dissociating from her hatred toward her parents, that hatred turned into a malevolent force that became an evil spirit within her.

Did Rachel hate her mother? My answer is an unequivocal yes. But she, too, is dissociated from her hatred, which comes through, then, as a hatred of herself. She has said, "I hate myself for that" (for not standing on her own feet). But how could she stand on her legs when they were cut out from under her? And without any legs to stand on, how could she express any strong anger against her mother? She was immobilized, frozen by fear and by her guilt and hate.

I don't believe that a person can fully surrender to love unless he can accept and express his hate. It becomes an evil force only when it is denied and projected upon innocent persons. To preach against hate is, in my opinion, futile. It is like telling an iceberg to melt with love. We need to understand the forces that create the negative emotions if we are to help people become free of them. To do that we must first accept the reality of these feelings and not judge them.

There is hate in all my patients and it has to be expressed. But first it has to be felt and recognized as the natural response to the betrayal of love. One has to sense how badly one has been hurt, psychologically and physically, to feel justified in expressing this feeling. When the patient feels this hurt deeply and is aware of the betrayal, I give them a towel to twist while they are lying on the bed. I suggest that as they twist the towel they look at it and say, "You really hated me, didn't you?" Once they can express that feeling, it is not difficult for them to say, "And I hate you too." In many cases it will come out spontaneously. Feeling such hate, one can mobilize a stronger anger in the hitting exercise. But no single expression in itself can transform the personality. Accepting the whole range of one's feelings, expressing them and gaining self-possession are the signposts along the road one travels on the voyage of self-discovery.

In this process of self-discovery the analysis of behavior and character is the compass that gives us the true direction. We have to understand the how and why of behavior before it can be changed. We must always start with the recognition and acceptance of a child's innocence. He has no knowledge of the complex psychological problems in the human personality. The love of a child for a parent, which is the counterpart of a parent's love for a child, is so rooted in nature that it requires a fair amount of sophistication on the part of the child to question it. Until that time the child will think that the abuse and lack of love are due to something he has done wrong. That is not a difficult conclusion to reach. For example, the conflicts between the parents are commonly projected onto the child. One parent will accuse another of being too lenient, which makes a child realize that it cannot please both. A child often becomes the symbol and the scapegoat of marital problems, and in many cases, although he is in the middle, the child is forced to take sides. I know of very few people who have emerged from childhood without a strong sense that something is wrong with them, that they are not what and how they should be. They can only imagine that if they were more loving, tried harder, and were more submissive, everything would be right. These people carry an attitude of trying to satisfy the other into their relationships, and are shocked to find that it doesn't work.

Healthy adult relationships are based on liberty and equality. Liberty denotes the right to express freely one's needs or desires;

equality means that each person is in the relationship for himself and not to serve the other. If a person can't speak up, he is not free; if he has to serve another, he is not equal. But too many people do not feel they have these rights. As children they were rebuked for demanding the fulfillment of their needs and desires; they were labeled selfish and inconsiderate. And they were made to feel guilty for putting their wishes ahead of those of their parents—like the patient who, as a child, complained to her mother that she was unhappy, and was told, "We are not here to be happy but to do what is required." She subsequently ended up being a mother to her mother. This is a fate that befalls many girls, robbing them of the right to fulfillment and joy. This betrayal of love by a parent must provoke a strong anger in the child against the parent, an anger which the child cannot express. The suppressed anger freezes the child's love, which turns into hate. This makes the child feel guilty and causes him to become submissive. Until these feelings of anger and hatred are released, the person cannot feel free and equal. Unreleased, they carry over into adult relationships.

Almost all relationships start with individuals being drawn together by positive feelings and pleasure. Unfortunately these things rarely continue to grow and deepen over the years. The pleasure fades, the positive feelings become negative and resentments build up, because without the feeling of being free and equal the individual feels unfulfilled and trapped. The suppressed anger is acted out in one form or another—either psychologically or physically—and the relationship is on the rocks. At this point the couple can break up or go for counseling in an effort to restore the good feelings they once had for each other. I have not seen many cases where counseling was effective. Most counseling aims to help the individuals understand each other and make a greater effort to get along together, but in effect it supports the neurotic attitude of trying. No amount of trying makes one more loving or more lovable. No amount of trying produces pleasure or joy. Love is a quality of being—being open—not of doing. One may earn a reward for trying, but love is not a reward. It is the excitement and pleasure that two people find with each other when they surrender to the attraction between them. Since all love relationships start with a surrender, their failure to continue stems from the fact that the surrender was conditional, not total, and was to the other person, not to the self. It is conditional upon the other person fulfilling one's needs and does not represent a full sharing of one's self. Some part of the self is held back, hidden, denied because of guilt, shame or fear. This held-back part, anger and hatred, is like a canker in the relationship, and slowly corrodes it. Removing the canker is the therapeutic task.

It is the existence of guilt, shame and fear in the unconscious that makes a person try. Diane, for example, was deeply ashamed of her sexuality, guilty about her anger toward her father whom she loved, and frightened that any expression of that anger would drive him away. She could not give herself freely and fully to a man because she did not fully possess herself. She was incomplete in her personhood and on some level she sensed her lack, which she then tried to compensate for by trying to serve and love. It only ended in her being abused. Of course she didn't merit the abuse; it is never deserved. It only happens to individuals who are in a dependent relationship. They become an easy object upon whom the other individual can vent his personal hostility, anger and frustration, which is derived from his early experiences with his parent. It is a law that the abused can easily become the abuser when a suitable object is available upon whom the suppressed hatred and anger can be acted out.

If we, as adults, look to another person for the fulfillment of our being—for happiness—we betray ourselves and we will be betrayed by that person. On the other hand, if we look to ourselves for the good feelings that are possible when we are in touch with ourselves and surrender to the body, we cannot be deceived and we will not be abused. We cannot be deceived because we are not dependent on another for our good feelings, and our self-respect will not allow us to accept abuse. With this attitude, all our relationships are positive, because if they aren't we end them. Individuals whose self-love and self-esteem are high are not lonely or alone. People are attracted to them because of their energy and the "good vibes" which radiate from them. Having self-respect, they command respect and are generally treated with respect. This is not to say that such persons do not get hurt in life. One cannot avoid pain or being hurt. But such individuals do not stay in situations in which they are continually being hurt.

Recognizing that joy is so desirable and the attitude of self-respect so positive, we must also bear in mind that they are not easy to realize. The surrender to the self and to the body is a very painful process at first because we get in touch with the pain that is in our bodies. Every chronic tension in the body is an area of potential pain that we would feel if we attempted to release the tension. Because of the pain one has to work slowly with the body. It is similar to the process of thawing out a frozen finger or toe. Too much heat too quickly applied results in a surge of blood into the area which bursts the contracted tissue cells and results in gangrene. The expansion of a contracted area, which is the equivalent of letting go, is not a one-shot affair. It is done little by little, over time, so that the tissues and the personality can become adjusted to a higher level of excitation and a greater freedom of movement and expression. But as slow as one works, pain is unavoidable because each step of expansion or growth entails an initial experience of pain that disappears as the relaxation or expansion becomes integrated into the personality.

The emotional pain is generally more difficult to accept and tolerate than the physical pain. The latter is localized, the former is pervasive. We feel the emotional pain in our whole body, in our being. The emotional pain is always the loss of love. One can be hurt emotionally in different ways—rejected, humiliated, negated, or verbally or physically attacked. But each of these traumas to the personality is, in effect, a loss of love. Being hurt physically by someone with whom one has no emotional connection results only in physical pain. One may be hurt physically all over the body but the pain is not heartfelt as emotional pain is. When a loving connection is broken we are cut off from a source of pleasurable excitement and life. The whole organism contracts, including the heart. There is a sense that one's life is threatened, which induces a feeling of fear. We survive this threat to our existence because not all loving connections have been cut. And, except for babies, people generally have a connection available to other creatures, to nature, to the universe, to God. Without some connection, I don't believe a human being can survive.

Individuals who have survived the loss of love in childhood have a big fear of breaking a connection. Some even say that a bad relationship is better than none at all. Just the thought of being alone is very frightening to many people. It awakens feelings that the person had as a child, when survival was tied to being part of a family. And it is related to the fact that being alone forces one to live intimately with the self. If one's self is weak, insecure and uncertain, to be alone with one's self is not enjoyable. But the insecurity that makes it difficult to live alone handicaps the individual in living with another person. One needs a connection to reduce the emotional pain, but it is never released through another person. One becomes more and more dependent. It ends with physical abuse that for some people seems preferable to the emotional pain of being alone.

Emotional pain is discharged by crying, which releases the state of chronic contraction in the body. To be effective the crying must be as deep as the pain, and must be connected with the conviction that it is hopeless to look for someone to restore the bliss of childhood, innocence and freedom. At the same time one must build a stronger self by energizing the body and sensing one's anger. A betrayed person normally would feel a murderous anger towards the betrayer. How does one deal with such anger when the betrayer is a parent? When the person betrayed is a child whose survival is dependent on that parent, the anger must be suppressed. But to suppress such a powerful feeling, an enormous tension must be built up in the body. That tension undermines the sense of self and cripples the individual's ability to be aggressive in the fulfillment of his needs. Without the ability to fight one becomes a victim who sees his goal as survival rather than joy.

I was consulted by a man in his late forties who complained about a feeling of tension around his waist and anxiety and uneasiness in his belly, which he had endured for many years. This man, whom I will call Harry, had many years of several kinds of therapy, including traditional psychoanalysis, but this problem was never treated. Harry was a strong, good-looking man who had a successful professional practice and, according to him, a good marriage. He was a physician, as his father had been before he retired. As a doctor, Harry was familiar with some of the literature dealing with body-mind problems. It disturbed him that his condition had not improved through his different therapies. He was familiar with Bioenergetic Analysis but had never experienced it. I was recommended to him as the authority.

When I looked at his body I was surprised how little feeling the whole lower part of it showed. Although normal in appearance, his legs appeared lifeless and weak. His buttocks were tightly contracted, with the result that his thighs and feet were rotated outward. I could see the band of tension in his lower back, but Harry felt no pain in that area. This deadness in the

lower half of his body was in sharp contrast to the seeming vitality in the upper half, which was well-developed muscularly.
As I pointed this out to Harry, he recognized the validity of my
observations. Although he had worked with other therapists on
a body level, no one had seen this disturbance, the meaning of
which was quite clear. Harry had been undermined by a strong
threat of castration anxiety which made him cut off feeling in
the lower part of his body.

To confirm this conclusion I asked Harry about his background. He was the youngest of three boys and, as the baby of
the family, was adored by his mother who had nurtured him
well. This created an enormous problem, since his father was
both jealous and angry about the feeling that the mother invested
in the boy. That anger was directed at Harry and took the form
of spankings whenever the boy did something out of line, such
as not doing what he was told, doing something he was told not
to do, or sometimes just speaking up against the parent. But
young children want their freedom to explore the world, and
will resist and rebel against restrictions; Harry's body bore wit-
ness to the extent of his punishment. Such punishment is easily
justified by the parent as being for the boy's benefit. He must
learn what is right and wrong and to take responsibility for his
actions. Harry did learn that; he was an obedient child and did
well in school, and his life moved forward in the properly ap-
proved channels. On the surface his life was successful, but deep
inside something bothered him and made him uneasy. He experienced it, however, only as a physical symptom and as a sense
that something was missing in his life.

During the discussion about his childhood and his relationship
with his parents, I brought up the issue of the oedipal conflict
that seemed so evident to me. Harry said that he knew about
the oedipal conflict and acknowledged that it was relevant to his
childhood situation, but he did not see any connection between

the oedipal conflict and his problem. Because he had no difficulty functioning sexually, he had no idea that he was psychologically castrated to a serious degree. He enjoyed his sexual relations with his wife; what was missing was passion. Harry operated from his head, not from his guts, which were tied up in fear of his father. Without passion there can be no joy.

Harry sensed that something was wrong, but he was unaware of the real nature of the problem, which can always be determined from the expression of the body and from a study of its form and motility.[4] The individual's problem is always manifested in his body since that is who he is. In Bioenergetic Analysis, the therapy always starts with an analysis of the body disturbance which is then correlated with the psychological problem that the person presents. Few persons are aware of how much their feelings and behavior are conditioned by the energy dynamics of the body. The first step in any integrated therapy —that is, one involving both the body and the mind—is to help the patient sense the tensions in his body and to understand their connections with his psychological problem. Harry came in with a physical complaint and had no awareness of its psychological implications. Most patients come in with a psychological problem and little or no awareness of its connection to the body. Harry accepted the psychological implications of his body problem when I pointed them out because of his previous therapeutic experience. But knowing about a problem or even gaining some insight doesn't generally produce a significant personality change. The passion that Harry needed to feel was not at the command of his mind; it was blocked off by the suppression of feeling and could be reawakened only when that suppression was lifted.

Harry had never fully expressed his anger at his father for the beatings he received. Those beatings broke his spirit and he be-

[4] Lowen, A., *The Language of the Body*.

came a "good boy," respecting his father and doing what was expected of him. He did not feel the injustice of his treatment, although in adult life he was very sensitive to political injustice. He did not feel any anger toward his mother for allowing the beatings and not protecting him against an angry and jealous father. His anger was bottled up in tensions in his upper back that he could not release because he had no ground to stand on. He had withdrawn his energy from the lower part of his body because he felt guilty about his sexual involvement with his mother. He was unaware of his guilt because he was out of touch with his anger.

Harry had to feel his loss before he would be able to mobilize the anger necessary to free his body. I started the therapy by having him do some bioenergetic exercises with his legs so he could sense the loss of feeling in them. The grounding exercise described earlier, in which the person touches the floor with his fingertips, proved helpful. After aligning his legs so that his feet were turned inward slightly and his knees positioned over the center of his feet, he could feel some vibration in his legs. Then, when I had him stand with his legs in the same position and his weight forward on the balls of his feet, he felt more in contact with his legs and felt more life in them, which helped him understand the direction of the therapy—namely, to "let down" into the lower part of his body. Over the bioenergetic stool his breathing was shallow and restricted to his chest, which was tight. He could not make a sustained sound that would allow the respiratory wave to go down into his belly, and he could not cry. He was aware that he held his feelings in and could not let go. And, of course, he didn't feel any anger. However, I could get Harry to kick while lying on the bed and to say "Leave me alone." It made sense to him to do this exercise and he had some feeling doing it. A specific sexual exercise proved very difficult for him and he could sense the pain and tension in his legs as he did it. The pain disappeared as soon as he stopped the exercise, which was unfortunate in one respect: Harry had to feel his pain a lot more intensely to evoke his suppressed anger. This is a general rule in therapy. A patient will react strongly only when his problem causes him sufficient emotional and physical pain to make his survival meaningless. For Harry survival meant being a good boy and doing what was expected of him. He hoped that this attitude would bring the reward of love, which has the promise of joy, but after much hard work Harry finally learned that joy is the feeling one has when one is true to one's self.

The beating of a child under any circumstances is physical abuse and should not be allowed. It gets results because the child is terrified, as any child would be who senses his impotence against the destructive power of a superior. If the superior is a parent upon whom the child is dependent, that fear becomes ingrained in the personality. When the child becomes an adult, two courses of action are open: The individual can take a passive position, hoping to win recognition and gain love by being good, doing good things for others, making few demands and causing no trouble. Harry belonged to this category. The other course is to be rebellious and act out the rage that is within. Such individuals become abusers of their children and spouses.

There are some who will swing between these two patterns depending on the situation. Neurotic patterns are maintained by the illusion that someone can provide the love that is sought so desperately. But no one can truly love these individuals since they are full of guilt and don't love themselves; it would be like pouring water into a sieve. It is hard to love someone who has no joy in his own being and so cannot respond to the love with joy. The failure of the relationship tends to make passive individuals more passive and angry ones more aggressive. By denying the betrayal, even though the denial is unconscious, the person betrays himself and sets himself up for a repetition of the childhood experience.

In some respects Harry's case was similar to Rachel's. She was physically abused by the same-sex parent just as Harry was, but where Rachel hated herself for her inability to be financially independent, Harry was very successful in his profession and took considerable pride in his position. His attitude to life was very positive in that he sincerely believed that with good will, one can attain all desired objectives. Therefore he felt no animosity toward his parents for the damage they did to him. He was also certain that he could overcome that damage by good will and effort. But with that attitude it would be impossible to reach the intensity of anger that could free his body from its debilitating tensions. He would have to fail in his therapeutic effort before he could feel the degree to which he was robbed of his manhood.

What could motivate a parent to spank a child so repeatedly that his spirit is broken? That was the meaning of the deep band of tension around the body at the level of the lower back which acted to split the body, separating the lower half with its sexuality from the upper half with its ego functions. But Harry was not a schizophrenic nor a double personality. He maintained his sanity and some integrity by abandoning his sexual nature. He could function sexually but on a mechanical level with no real passion. There was no passion in any aspect of his life, including his work. Did Harry's father hate him? Did Harry hate his father? To both of these questions I would say yes. But what about his feeling toward his mother, who put him in the position of being a rival to his father for her love? Or for not protecting him against his father's anger? His relation to her is complex. Through her seduction, she made him feel special and superior but it was at the expense of his sexuality and was a way to bind him to her. His guilt about his sexual feelings for his mother was as great as his suppressed anger and hatred. Because of this guilt he could not see his father as the cold, sadistic person he was. And because of that guilt he could not surrender to love.

I was consulted by Louise, a woman therapist who some years ago was tormented by a feeling of guilt over the suicide of one of her male clients. She was well aware that she was not responsible for his death, but she felt that she should have paid more attention to his expressions of distress, which would have indicated suicidal thoughts. She felt she might have been able to do more to alleviate his distress and prevent the suicide. Even recognizing that she was a competent therapist and had acted responsibly, she couldn't free herself from the torment of guilt.

This patient described herself as meek and unaggressive. In the course of her previous therapies she had made some improvement in her ability to be self-assertive. In this study I have consistently emphasized that guilt is directly connected to the suppression of anger. That suppression undermines the body's good feelings. In their place, one senses a disturbing element which feels bad. The feeling that something is wrong or bad is the basis for the sense of guilt. One cannot feel guilty when one feels good in one's self. Superimposed on the feeling of something being wrong is a judgment on the self that one should do more, try harder, be more responsible for others. Louise was raised with these commandments.

As we explored her history she told me a story of physical abuse that was shocking. When she was young she was spanked regularly by her father with his belt or his hand, often on her bare bottom. He was a violent man and she was terrified of him. In her past therapies she had expressed some angry feelings toward him but never with the intensity one should feel about such abuse. I asked her if she had had any death wishes for her father. She said no. I was sure, however, that she had a tremendous rage within her toward her father for his vicious treatment of her, a rage she had suppressed out of fear. Her feeling of guilt stemmed directly from this suppression and was transferred to her patient, whom she was unconsciously trying to save from her own anger at men.

I did an exercise with Louise to help her sense her rage. This exercise has been described in Chapter 5 but I will repeat it here because it is so helpful in getting a patient to feel his anger. I had her sit in a chair facing me as I sat in another chair three feet from her. I asked her to make two fists, thrust her lower jaw out, open her eyes wide, shake her fists at me and say, "I could kill you." It took several tries before she yielded to the exercise. When she did, her look was maniacal and she could feel the overwhelming intensity of her rage. I have done this exercise many times with individual patients and in groups, and no one has ever actually been attacked. In this exercise the expression is one of anger, not rage, because the person is never out of control. But in almost every case it gives the person a sense of strength and power and a stronger feeling of the self.

Following the exercise Louise lost her meek look. Her face was more alive and stronger looking. She understood the connection between her anger at her father and her guilt for the suicide of her client. And she felt greatly relieved.

When a woman suppresses her anger against her father for his betrayal of love, it is transferred to all men even though it is not consciously acted out. It will come through in subtle ways to destroy the relationship. Similarly, men who have suppressed their anger at their mothers who have dominated them or failed to protect them against a hostile father will necessarily project that anger upon all women. Each woman stands for the seductive and, at the same time, castrating mother. Until that anger is expressed the man does not feel free to be himself, with the result that his relationship with women is handicapped. The partner will be blamed for the lack of fulfillment in the relationship that actually stems from a sense of unfulfillment in one's person. Blaming the partner is a betrayal of the love one was given. To make a love relationship work one has to bring a feeling of joy to it; this requires that one be free from guilt so that one can express all feelings directly and appropriately. For that one has to know one's self deeply, which is the goal of therapy.

CHAPTER 8

SEXUAL
ABUSE

Sexual abuse is a most heinous form of the betrayal of love since sexuality is normally an expression of love. The abuser approaches his victim as if he or she were offering love, but then takes advantage of the innocence and/or helplessness of the other for his personal need. It is the betrayal of trust which constitutes the most damaging aspect of this crime, but the physical violation adds an important dimension of fear and pain to this destructive action. Individuals who have been subjected to sexual abuse generally bear the scars of their experience for a lifetime. Most serious is the victim's suppression of the experience because of feelings of shame and disgust about what happened. But when these feelings are suppressed, it leaves the individual with a deep sense of inner emptiness and confusion. Victims of sexual abuse cannot surrender to their body or to love, which means that there is no chance of fulfillment in their lives. For them the voyage of self-discovery is a most frightening venture. Their treatment requires a special awareness of this problem.

How common is sexual abuse? That depends on what we

consider to be sexual abuse. Statistical studies based on question-
naires sent to adults indicate that 30 to 50 percent of those responding reported that they had been abused as children. If any
violation of a child's privacy regarding his body and sexuality is
regarded as sexual abuse, the incidence, I believe, could be greater
than 90 percent. A female patient recalled her feeling of shame
and humiliation when, at the age of three, she was made by her
family to pose nude for a photograph. Public comments about a
child's developing sexuality can well be considered a form of
sexual abuse. When a father spanks his young daughter on her
bare bottom, it is, in my opinion, an act of sexual abuse as much
as physical abuse. If the father derives a sexual excitement from
his actions, the child senses it. One patient told me that she asked
her husband to spank her bare bottom—something which excited her sexually to a point where the sexual intercourse that
followed was the best she had experienced. This is typical masochistic behavior.[1] It stemmed, no doubt, from the fact that this
woman was spanked by her father when she was a child in a
similar fashion, which greatly excited her sexually. Masochistic
or sadistic practices associated with sex stem from childhood experiences which have become imprinted in the child's personality.
Many women use masochistic fantasies, such as being tied up
while they are engaged in the sexual act, to help them reach a
climax. I would go so far as to say that any beating of a child
by an adult has sexual implications.

Today, however, we are aware that many cases of sexual abuse
involve direct sexual contact between an adult or adolescent and
a child. We also speak of such cases as a form of incest. Where
such direct contact occurs, it has a very destructive effect on the
child's personality, the severity of which is inversely related to

---

[1] In this regard see Reich's description of masochism in his book, *Character
Analysis*.

the age of the child—that is, the younger the child, the more severe the damage. I was shocked to learn of cases where the child was an infant. When sexual abuse happens at a very young age, the child represses all memory of the events by suppressing the feelings associated with them. Suppression involves deadening a part of the body. When the feelings come alive again, memory is awakened. This is illustrated in the following case.

Madeline was almost fifty years old when she first became aware that she had been abused as a very young child. She sensed that something was amiss in her life because in both of her marriages she was physically abused by her husbands. However, she did not connect the abusive treatment by her husbands to the possibility of sexual abuse in very early life. Both her parents had been alcoholics and the family was dysfunctional, but since the family was secretive and kept Madeline away from other children, she saw her disturbed family life as the way life normally is.

Madeline was a survivor. She ran a successful business and had raised four children to adulthood. She also had the courage to leave the two men who mistreated her, but she felt no real anger toward them. She only knew that she had to get out of the relationships. One day, Madeline's closest woman friend encouraged her to join an incest survivor group, which she did. When she heard other women tell about being sexually abused by a parent as children, it dawned on her that she must have suffered a similar experience. The idea terrified her but it would not go away. She began to feel the fear in her body, which she could then associate with an act of abuse.

Madeline came to me following an experience with a male therapist who, at the end of a session, gave her a hug pressing his pelvis against her. She was both angry and frightened. After relating this incident she went on to tell me about the awareness she had gained in the incest survivor group that her father had used her sexually when she was one year old. That information struck me as unbelievable, but since I had no reason to question her feelings, I accepted it as possible. Over the next two years of therapy I became convinced that this had happened. As the work with her body progressed and she began to have sensation in her pelvic floor and rectum, she panicked. The fear was so great that she would cut off all feeling and withdraw from her body. This fear confirmed her belief that she had been penetrated rectally as a very young child.

The phenomenon of withdrawal from the body is a dissociative process typical of the schizoid state, in which the conscious mind is not identified with bodily events. The "I" of the conscious mind acts as an observer of what is happening in the body. The subjective sensation of being the person in the body who is experiencing the event or action is missing. The connection between the observing I and the acting I is broken. The reason for the break is that the experience is too frightening to be integrated by the ego, which protects itself by splitting off from the experience. In extreme cases where the fear amounts to terror, there is a more total break of the connection with the body, resulting in a condition of depersonalization which is characterized as a nervous breakdown that could lead to schizophrenia. Madeline never became schizophrenic. However, the connection in Madeline between mind and body was vulnerable, subject to being broken whenever her fear approached the level of terror. This constituted the schizoid state.[2] In an advanced state she would withdraw from her body to the point where she didn't feel she had a body. Fortunately this severe state of dissociation was shortlived. She would slowly restore the connection between the conscious mind and the body sufficiently to feel the reality of the

---

[2] See Lowen, Alexander, *Betrayal of the Body* (New York: Macmillan Publishing Company, 1970) for an in-depth analysis of this condition.

physical self. But that connection was superficial rather than deep, preventing her from sensing how damaged she had been. Below the surface she was a terrified child.

In her daily life one would not think of Madeline as being such a terrified person. She was intelligent and she could handle the ordinary events of her life reasonably well. The terror arose only when a strong feeling tended to surface and throw her out of control. Since she needed to be more aggressive to protect herself against abuse, I had her do the exercise of kicking the bed and saying loudly, "Leave me alone." If her voice rose to where a scream broke through, she would curl up in a fetal position in the corner of the bed, whimpering like a little child in terror. It would take a number of minutes before the fear subsided sufficiently so that she could return to enough of her "normal" self so she could leave my office with some sense of sanity. It was also very difficult for her to cry, because any break-down of her control threw her into terror. I believe it was my sympathy, support and encouragement to open up her anger that enabled her to experience a strong anger without becoming ter-rified and dissociating.

Observing Madeline's daily functioning, one would not suspect the degree of disturbance in her personality. She operated from her head with very little body feeling. However, she did have sexual feelings and many men were attracted to her. She claimed that she enjoyed her contact with them, which I believe was true, but it was a dissociated experience in that she was not connected to her sexuality, which was limited to the genital apparatus and was without passion. On the superficial level of her personality she was a mature woman, but on a deep level, she was a terrified child, lost and helpless. The seemingly mature woman was just on the surface. At any real depth of feeling one encountered the frightened child. As she made more contact with the terrified child, she began to feel her body in a different way—not as something she could use but as the person she was. And her terror and fear diminished.

Given the horror of Madeline's infancy and the resulting disturbance in her personality, one might have difficulty in comprehending her sexual pleasure. One must realize, however, that she was a split personality and that her sexuality was very superficial, like her other feelings. She could not connect to her sexuality as an expression of her self any more than I could connect to the scream that had issued from my throat during my first session with Reich. A scream is an intense sound but there was no feeling of intensity in me. Similarly, sex should be an intense experience, but for Madeline and other sexually abused individuals, it is not experienced as such. Any abuse of a child, physical or sexual, which terrifies the child, causes him to dissociate from his body. It was difficult for Madeline to experience any intense feeling without becoming terrified and cutting off from her body. Her body could not tolerate the charge and her mind could not integrate the emotion.

In the therapy, Madeline worked physically to deepen her breathing and to give in to feeling in her body. However, each step deeper into a stronger feeling threw her into an episode of terror in which she closed up and withdrew from her body. While she would regain self-control following a session in which stronger feelings emerged, she told me that she would be out of her body for some time afterward. Being out of the body meant cutting off all feeling and operating solely from the conscious mind. Slowly the fear diminished and she could stand more emotion and more feeling in her body without becoming terrified and cutting off. If a session was very strong for her and she left her body, she could now return rather quickly, which she recognized as significant progress. I recall the session in which Madeline remarked with excitement, "I can feel my feet."

The issue of the very early abuse remained a very difficult one to resolve. She felt extremely vulnerable in and around her anus, and one had to wonder how she could have seemingly normal sexual relations given the amount of fear she felt in her pelvic floor. But Madeline told me that she enjoyed sex even with men who abused her. In fact, she was quite seductive, although she was not fully conscious of that aspect of her behavior. Although she was a terrified little girl on a deep level, she was also a sophisticated woman on the surface who was excited by and welcomed the sexual attentions of men. Sophistication is the right word, for while it denotes the absence of innocence, it also indicates a lack of guilt, which is unreal. To survive, Madeline had accepted the perversity of her world as normal. If sex is what the world is about, she would learn how to use it. Thus, despite the sexual abuse of her childhood and the physical abuse of her married life, Madeline felt no hatred of men and no anger toward them.

Both hatred and anger were in her, but these feelings were cut off by the need to survive, which she did being sexually available to men. After all, if they are so desperate for sexual contact and release, why not give in to them. Submission removes the threat of force and violence and denies the fear. That no man would hurt a woman if she gives in to him is the false reasoning of women who have been abused.

There is, however, another element in the personality of the woman who was abused as a child that shapes her behavior as strongly as the fear and helplessness associated with the abuse. That element is a strong sexual excitement limited to the genital apparatus and dissociated from the conscious personality. The early sexual abuse both frightens and excites the child. It is not an excitement that can be integrated by the immature body and ego of the child, but it makes an indelible imprint on the body and the mind. The child momentarily enters the adult world, which shatters its innocence, but from that moment sexuality becomes an irresistible and overwhelming but split-off force in the personality. Marilyn Monroe is an example of this condition. She embodied sexuality but she was not a sexual person. It was as if she acted a sexual part without being identified with it on an adult level. Her adult personality was split between a sophisticated mind and a childlike dependency and fear. She was sexually sophisticated but it was very superficial and covered an underlying sense of being lost, helpless and frightened. In a previous study I had characterized Marilyn Monroe as an example of a multiple personality.[3]

A young woman was referred to me to help her understand her confused life. Betty, as I will call her, had been brought up in foster homes and related a history of sexual abuse from the time she was ten years old. Her confusion related to the problems she had with men. They were attracted to her (she was an attractive woman) but relationships with them went nowhere. The remarkable thing about this person was that she exuded a sexual aroma which was almost palpable in the room. Since it was her natural odor, she was completely unconscious of it. Like Madeline, she existed on two levels: one superficial, on which she functioned as a sexual woman—sophisticated and competent in worldly affairs; the other a deeper level, on which she was a terrified little child unable to cry deeply or become strongly angry. She behaved as if she was possessed by a sexual charge which was an alien force in her personality, one over which she had no control. Betty wasn't conscious of the effect on a male of this strong sexual aroma since she did not perceive it. It was not there all the time, but probably only when she was unconsciously trying to seduce a man into a sexual intimacy. Her seduction, however, was not an expression of passion but of need.

[3] Lowen, Alexander, *The Spirituality of the Body* (New York: Macmillan Publishing Company, 1990).

Betty needed my help and one way to get it was to excite my sexual interest in her through her sexual emanation. She was not consciously giving off a genital aroma; it was caused by the fact that her vagina was charged and excited, which, however, she did not feel. That excitation stemmed from the abuse and was not her own feeling, thus she did not identify with it. She had learned to use it early in life through her experiences in foster homes. She had found that while her foster mothers were hostile to her because as women they distrusted her sexuality, men responded to her sexually. As a child, consciously or unconsciously, she tried to get some support from her foster fathers but they used her for their own ends. I am sure that on some level they felt sorry for her and wanted to help, but on a more immediate level they took advantage of her need and helplessness to use and abuse her sexually. In her desperation she acquiesced, believing on some level that they loved her. It didn't work. The foster mothers were aware of what was going on and Betty would be moved to another foster home, where the same events would take place.

Betty did not stay long in therapy with me and I did not have the time to analyze her background fully. She had repressed most of her early memories, and at the time I saw her many years ago, I did not have the depth of understanding about these issues that I have now. One learns from one's failures. But I had an intuition that this must have been the situation since Betty was referred to me by the man she was living with and working for, whom I knew had the personality of an abuser. His approach to a woman was to offer help which he believed was genuine, but when she responded he used her sexually. This is the kind of man Betty would be attracted to through the operation of the compulsion to repeat. Madeline was acting through a similar compulsion to become involved with men who physically abused her, as, also, was Martha, whose case I presented in Chapter 3.

As long as these women are hung up on their search for a man who will love them and protect them, they will be used and abused. Their relationships with men cannot work out. Men respond to them as sexual objects, not as sexual persons, because the women do not see themselves as persons. Their sense of selfhood is too badly damaged by the sexual abuse.

Sexual abuse has the effect of prematurely overexciting the sexual apparatus of the victim. Despite the fear they feel, the sexual excitement of the contact becomes imprinted in the personality because it remains undischarged in the sexual apparatus. Their attraction is to a man who is seen as similar in personality to the abuser and their sexual submission is an unconscious attempt to free themselves from their hang-up by reliving the situation and completing the discharge. Only it never happens because of the dissociation.

Lucille told me that she was constantly aware of an excitement in her vagina which she experienced as a foreign or alien element. A major part of her sexual actions were aimed at discharging this excitement so that she would be free from its torment. That didn't work because the freedom she felt after sex was short-lived. She was literally possessed by an alien force, the sexual charge of her abuser, which she was unable to discharge. Discharge occurs only when the excitation flows downward through the body into and then out of the genital apparatus. Violation at a young age, that is, before the ability to discharge the excitation through orgasm develops, causes those organs to become charged with a force over which the individual has no control. The young girl has literally been dispossessed from her own genital organs.

The victim of sexual abuse can repossess her sexual organs by allowing her excitation to flow downward and into them. This is the normal sexual pattern but in these cases it is blocked physically by a band of tension around the waist and psychologically by strong feelings of shame about one's sexual parts, which are regarded as unclean. So many women feel the shame of their sexuality because it was not allowed to develop as an expression of love. And yet, sexuality is an expression of love, a desire to be close and united with another person. Unfortunately, that love is often mixed with its opposite—hostility. Most human beings have ambivalent feelings because of their childhood experiences, in which the love of their parents was crossed with negative and hostile feelings. That is very evident in the cases I've already described, but I believe that it is also true of most family relationships. One cannot surrender fully to love after having been betrayed by those one loved and trusted. I have seen a number of sexually abused women with similar patterns of behavior. These are intelligent women whose lives have been severely damaged by their victimization. All have multiple personalities stemming from the conflict between their sexual excitement and their fear, between a sense of being desirable and a strong feeling of shame. And, in all cases, sexuality was not an integrated aspect of their personality.

Some years ago I was consulted by a very beautiful woman in her forties named Ann, whose problem was an extreme rigidity of her body which made all her movements difficult. She related (as an example) that when she was chosen queen of the annual festival at her college, she had trouble walking down the steps to receive the crown. Doctors had been unable to treat the condition because they found no neurological disturbance. She believed that it had an emotional basis. From the ages of twelve to eighteen her father had had regular sex with her. He was in love with her and she was in love with him. She described him as an outstanding man in his community, looked up to and admired by everyone, including his daughter. Her explanation of her condition was that she could not allow herself to reach a climax in their sexual contact because then she would feel guilty and ashamed. By not surrendering to her sexual feelings she could believe that she was doing it for her father—that he needed her. She claimed that she loved him, which I am sure was true. And I am sure that he loved her—but he also betrayed her.

His betrayal had made it very difficult for her to surrender to any man sexually. She had been married for a long time to a man she loved, but she said it took many years before she was able to have an orgasm with him. Given the degree of bodily rigidity from which she still suffered, I didn't think that she could easily or fully yield to the passion of love. She had been far more damaged than she knew or acknowledged.

Betrayal, like treason, has always been regarded as a capital offense, one that, in olden times, merited punishment by death. I am sure that Ann held back a tremendous anger toward her father for his behavior. Her rigidity was not just a means to control her passion; it also served to suppress and control his rage. For just as we melt with love, we stiffen and become cold with hatred. The hatred, however, was in the outer layer of muscles, not in her heart. Like all sexually abused individuals, she was split: in her heart she loved her father but in the muscular layer she resisted and hated him. Her beauty was an expression of her sexual attractiveness, but her sexuality was not fully available to her.

I only saw Ann on two occasions because she lived in a different part of the country. As we spoke about her life and her problems I felt that she was not prepared for nor desirous of opening up her anger toward her father. But without the release of this anger it would be impossible to soften the muscular rigidity which bound her as if in a strait-jacket. In individuals who've been sexually abused there is a strong resistance to the venting of anger against the abuser. In part that resistance stems from a feeling of guilt for having participated in the sexual acts, whether that participation was voluntary or forced. But it also stems from the fear of the anger itself, which is experienced as a desire to kill. Killing a parent is the most heinous crime and yet the betrayal was by the parent. The resolution of the conflicts created by sexual abuse can only come through a therapeutic program that provides a controlled situation for the expression of such anger.

Studies have shown that male children have been sexually abused almost as much as females. Some have been violated by their father, some by another older man, some by older siblings. When this happens it has the same effect upon a boy's personality as it does upon a girl's. If there is anal penetration, the child can experience intense pain and fear which can cause him to dissociate from his body as Madeline did. The sexual abuse of a boy by an older male undermines the child's developing masculinity and makes him feel ashamed and humiliated. I don't believe that such experiences create a homosexual tendency in the boy's personality, but the resulting weakness in the boy's masculine identification could predispose him to that pattern of sexual behavior.[4] The damage to a child's personality is caused by the emotional impact of the experience. Fear, shame and humiliation are devastating feelings to a child who has no way to release and recover from the insult of this trauma. The physical abuse of a child by his father, as in repeated spankings, has a similar effect upon the boy's personality, and as I pointed out in the preceding chapter must be regarded as a form of sexual abuse.

Sexual abuse is as much an expression of power as it is an expression of sexual love. The sense of having power over another person acts as an antidote to that feeling of humiliation which the abuser suffered when he was abused as a child. The issue of power also enters into sexual activity even when it occurs between consenting adults, as in sadomasochistic practices. The

[4] See Lowen, Alexander, *Love and Orgasm* (New York: Macmillan Publishing Company, 1965) for an in-depth analysis of the homosexual personality.

abuser generally is an individual who feels incapable of being a man or a woman on a mature level. That feeling of impotence disappears when the victim is a child, a helpless adult or a submissive partner. The abuser feels powerful in this situation, which means that he also feels sexually potent. When feelings of power intrude into a sexual relationship, it always turns into an abusive one. A man who needs to feel power to be sexually potent will necessarily abuse a woman. Often when the man is seen as powerful the woman becomes excited and more able to surrender to him. This, of course, is only true of women who have been victimized and feel impotent themselves. Diane, whose case I reported in an earlier chapter, remarked that the best sex she had experienced was with the husband who physically abused her. Abusive behavior between adults denotes a sadomasochistic relationship that allows the individual to surrender to his sexual excitement. For the sadistic partner it is the feeling of power over the other, manifested in actions intended to hurt or humiliate the partner. For the masochist, submission to the pain and the humiliation removes, temporarily, the guilt which blocks the sexual surrender. In the submission, guilt is transferred to the abuser, allowing the victim to pretend innocence.

On one level, abusive behavior expresses hatred—the desire to hurt another. But we must also recognize that there is also an element of love in the relationship. Reich recognized the connection between sadism and love, for he believed that a sadistic action originates in a desire for contact and closeness with another. It starts out as an impulse of love in the heart, but as that impulse moves to the surface, it is twisted by tensions in the musculature related to suppressed anger, turning it into a hurtful act. The victim may sense this dynamic, especially when the abuser is a parent acting out on his or her child. I am suggesting that a young child who is extremely sensitive to the emotional nuances of behavior may recognize that the punishment or abuse is intended to be an act of love. The love turns sadistic when it cannot be expressed. This recognition could prevent the child from feeling the fullness of its anger against the abuser. The child also recognizes the pain in the abuser which prevents that person from expressing love easily and freely. One, then, feels sorry for the abuser and identifies with him.

Young male children are physically abused not only by their fathers but by their mothers. In Chapter 4 we saw a case in which the mother's physical abuse of her son was intended, consciously and unconsciously, to break his spirit and to make him submissive to her. No child can stand up to his mother's or father's violence. Anyone would inevitably be broken by such an experience. But the break is very rarely total, for that would be death (although we know that such cases have occurred). In the depth of the child's body a core of resistance remains to support life and provide some sense of identity. The strength of that core depends on how the parent relates to the child after the abuse. Having discharged her suppressed rage, a mother, for example, may feel a deep love for the child she has just abused. To the degree that the child feels that love, the damaging effect of the abuse is partially reduced. If the child feels a deep hostility from the mother, which amounts to a cold rejection of the child, he could become schizophrenic. Children are aware on some level that the beating or physical abuse is preferable to a cold rejection, which is emotional death.

The general proposition that individuals tend to act out on others what was done to them helps us understand a mother's seemingly irrational behavior toward her children. If she was humiliated as a child for any sexual expression, she will tend to do the same to her children. This tendency to act out on helpless inferiors can only be avoided if the individual is acutely conscious of what was done to him and keenly aware of its destructive effect upon his personality and his life. This awareness implies that one can sense an anger against the parent for the abuse and violation. A mother who is ashamed of her sexual feelings will shame her daughter for any expression of such feeling. Mothers identify with their daughters and project upon them the negative aspects of their own personality. Thus a mother can see her daughter's sexual behavior as whorish because that was how she was seen as a child. By criticizing her daughter for being sexual, she, in effect, is saying, "You are the bad one, the dirty one. I am clean." On the other hand she can also project upon her daughter her unfulfilled sexual desires, unconsciously wanting her daughter to act them out so that the mother can obtain a vicarious thrill from her daughter's actions. Actually, both attitudes can exist in a mother, one consciously degrading the girl for being sexual, the other unconsciously pushing her to act out sexually. This unconscious identification on a sexual level of a mother with her daughter has a homosexual aspect. The failure to see this aspect of the parent's relation to the patient can become an obstacle that prevents the patient's movement toward independence and fulfillment.

The abused becomes the abuser by an unconscious identification with the abuser. This is the other side of the coin to be recognized and accepted by the patient for full self-acceptance to occur. Rachel reported a dream in which she saw a young girl standing next to her, one side of whose face was red as if it had been rubbed against something. She realized immediately in the dream that she had rubbed the girl's face against her own pubic area. She was horrified to think that she could do such a thing. But in the dream, Rachel was also the girl who was violated. If the dream reflects an incident that may have happened to her when she was a child, why would she want to act it out against someone else? Acting out on others the abuse she suffered allows her to feel "I am not alone in my shame." But there is another motivation for such behavior: When a child is abused sexually it is both exciting and frightening to the child. Every young child is fascinated by the genitals of his parents. For one thing, they were the source of his life. They were also the keys to his underworld of secret pleasures and fears. But because of the fear, the abuse and the accompanying excitement are suppressed, and only their imprint remains. The person is strongly drawn to repeating the experience, often as the abuser, but also as the abused. I believe this is how an adult becomes obsessed with sex with children. His libidinal development is handicapped because some of its energy and excitement is encapsulated in the repressed memory and associated feelings. Bringing these incidents to consciousness is the first step in releasing the bound energy. Bringing the buried experience to light reduces the shame, allowing the individual to feel his hurt and his fear. Accepting both feelings would allow him to cry, releasing the pain, and to become angry, which would restore his integrity. But that anger has to be real and intense to purify and liberate the spirit.

Mothers are in a unique position to act out sexually on their children because they are more involved with the child's body than fathers are. How they touch the child's body can have sexual implications, just as the fear of touching does—namely, that touching can arouse sexual feelings. One mother remarked about her two-year-old son, "His penis is so cute, I could take it in my mouth." The feeling behind that statement must somehow be communicated to the child when his genitals are exposed. His sense of privacy about that organ is gone. Her feeling invades his pelvis and possesses his genitals. It is not just any looking at a child's genitals that will disturb him, it is looking with some sexual interest or awareness. Mothers are often advised to clean a little boy's penis to avoid possible infection. I don't believe that is a good idea and I don't believe it is necessary. Since time immemorial boys have grown up without needing such intervention. The danger in all parent–child relationships is that the bonding will have a strong sexual element. This element will be denied and suppressed by both parent and child, but its effect on the latter is devastating, as the following cases show.

Max was an only child whose father had died when he was young. He was raised by his mother, whom he described as a powerful woman afraid of no one. Max, a man in his thirties, was a psychologist and familiar with Bioenergetic Analysis. He recognized that he had many tensions in his body which blocked him from experiencing any pleasure or joy in his life. He worked hard but had no real satisfaction from it. He was always pushing himself, trying to achieve a position that would allow him to feel relaxed and easy in his life, but nothing went easy or right. He felt that he had to fight for everything he wanted, and this attitude involved him in several lawsuits. The same problems and difficulties arose in his relationship with his wife. They had constant little fights which resolved nothing, since his problem was a personal one. Knowing Max, one would describe him as a tormented man, but while he recognized his torment, he did not know what caused it.

Physically he was good-looking, strong and energetic, but his body had a chaotic quality. When he breathed, the respiratory waves had a jerky, convulsive look and did not flow easily. The problem was most noticeable in the lower half of his body. His pelvis was held tightly and did not move with his breathing. His legs, though they were developed muscularly, did not give him any sense of support. In a grounding exercise, they shook rather than vibrated, and collapsed on him. Lacking a sense of support from the ground, he held himself up with his head, always thinking, calculating, maneuvering. He experienced this manner of life as very frustrating.

In the first two years of therapy Max made little progress. He struggled, he pushed himself, and he tried, but could not break through to any strong feeling. It was almost impossible for him to surrender to his body. However, his resistance was unconscious and I could only point it out to him. He was discouraged and stopped coming to his sessions. I did not encourage him to continue, because the last thing he needed was to be pushed by me. I had no feeling that further effort on his part or mine would help. During the sessions he had focused on his relationship with his mother who was still very much involved in his life and still trying to control him. He rebelled but could not free himself, though with my encouragement he had slowly distanced himself from her.

He resumed the therapy about a year later. He sensed that I understood his problem even though I had been unable to help him achieve the change he wanted. However, in some respects his attitude and his life had changed. He was less pushy and less belligerent. His relationship with his wife was better. He had continued to do the bioenergetic exercises at home, mainly kicking on a bed and breathing using the bioenergetic stool, and he sensed that they helped him feel better. I sensed a change in him—he was more open to the idea of letting go. Over the stool he could give in more to his sadness and cry, though it didn't go deep enough. His kicking was stronger; it was focused on his desire to be free from his mother and from the pressure she had put on him to fight the world. At around this time two major events occurred which promoted his bid for freedom. The first was the death of his mother. In some deep way it freed him from her influence. The second was the birth of his first child, which both he and his wife had desperately wanted. The child brought light and joy into their lives. While this was helpful to Max, what he needed was to feel these qualities in his own body.

Surrender is simply a "letting go" to one's feelings. The first big breakthrough for Max occurred when he was kicking the bed using the words "Leave me alone," which he felt were addressed to his mother. When a patient talks about wanting to be free from pressure, I suggest that he demand it. A person has to be prepared to fight for what he wants if he is to get it. Max's belligerence was not that of a fighter; it was more manipulative than confronting. His underlying rage was so great that he dared not let it come out fully or freely. No matter how much awareness he had of this problem through our discussions, it didn't release him from his fear, and it couldn't unless he could fully express his feelings of protest and anger. The exercise of kicking the bed is ideal for this purpose because there is no danger attendant on letting go of control. The patient will not hurt himself or anyone else and the foam mattress will not break. To reach the breakthrough point the kicking has to become spontaneous and the voice has to reach the level of a scream. It happened for Max. He "let go" to the exercise and the feeling took over. As soon as he finished he felt a difference: lighter and freer. Of course, that gain had to be consolidated and developed by repeating the exercise in succeeding sessions, giving Max an increasing sense of himself as a free person.

However, the breakthrough did not extend to the lower part of his body. It did not free up his pelvis. This next step in his therapy required work with the legs and pelvis. For that purpose I relied mainly on a falling exercise that charges the legs and feet energetically, allowing the charge to break through into the pelvis and free it up. In this exercise the patient stands in front of the stool with his back to it. With his hands on the stool behind him to hold his balance, he bends his knees until his heels are raised off the floor. The weight of the body is entirely on the balls of the feet, but the patient holds against falling forward by pressing down with the heel even if it is kept off the floor.[5] The patient is instructed to keep the charge in his feet and not let himself fall. When this exercise is done correctly, the pelvis will spontaneously move with the respiration. It didn't happen for Max although he tried the exercise many times in a number of sessions. But each time he did the exercise more feeling came through in his legs.

Max could not maintain the position for more than a minute. His knees would collapse and he would fall to the floor. As we discussed this problem, he said, "I can't stand up to her. She's all over me, smothering me." Saying this he became very angry and drove his pelvis forward with the words, "Fuck you!" After saying this his pelvis began to move freely with his breathing. It was the breakthrough he needed.

Whether his mother had actually lain on his body when he was a child, he did not know. It is likely that as she lay in bed with him when he was a small child, her body was pressed against him exciting him sexually. That she was sexually involved with him could not be doubted, and after this experience he did not doubt it. His body reflected the fact that he had been highly excited by his mother sexually but was unable to get away or release the charge. The torment almost drove him crazy. He did not dissociate from his body as Madeline had done because he was not terrified of his mother. She did not hate him and she had not violated him physically. Rather she loved him, but it was too much, too sexual. In focusing her sexual love on Max, she used him to fulfill her romantic dream, but for Max it was a form of sexual abuse.

[5] See Lowen, Alexander, *Bioenergetics* (New York: Coward, McCann & Geoghegan, Inc., 1975; Arkana, 1994) for an explanation of energy dynamics in the body when doing this exercise. The use of a falling exercise to allow a surrender to occur is discussed in my book *Fear of Life*.

Robert was another man who was bonded with his mother. He was a bright, attractive man who couldn't find any fulfillment in his life or in the world. He wanted to do an important work but he couldn't. He wanted a deep relationship with a woman but it didn't happen. On one level Robert felt that he was special, but on a deeper level he felt insecure and frightened. The excitation did not flow freely through his body and he had marked tensions about the pelvis which reduced his sexual charge. His sense of specialness was evident in the way he related to people. I would describe him as a charmer. He knew what to say and how to say it, which denoted a high degree of ego control. But as a result, Robert was afraid and unable to surrender to his body, to himself and to life.

He described his background as follows: "My mother's energy was very frenetic. On one hand it was very exciting but on the other it was overwhelming. When I was caught up in it, I would lose complete sense of myself.

"My brother was very jealous of her involvement with me. He was three years older and twice my size. He beat up on me and tortured me physically and psychologically. He had no trouble expressing his anger. I looked up to him and feared him.

"I sense that I made a political deal with my mother but I had to surrender my self. She convinced herself that I was perfect, could do no wrong and never lied. At the same time I lied all the time just as she did. But in our alliance these lies were overlooked. It was an acceptance of mutual corruption. I saw her as perfect. I identified with her."

Robert could have become a homosexual. What saved him was some identification with his father, who tried to intervene for him with his mother. When she threatened the father, however, he withdrew, and eventually turned against Robert, which allowed his mother to possess him.

The effect on the boy was very destructive. Robert said, "I felt

I was on the brink of going crazy, going right into madness. I used to examine her closet and drawers, handling her lingerie. I couldn't stop the urge and I couldn't contain the excitement; then, as I reached puberty and became involved with my friends, it subsided."

As an adult Robert was able to move out into the world and try to establish a life that would be fulfilling, but it was not easy given the degree of disturbance in his personality. He did survive, which meant that he became one of countless young people who are striving for success but whose lives provide no real feeling of joy or fulfillment. Those who enter therapy are the fortunate ones, for they have a chance to work through their problems and hang-ups and find life's real meaning. It is not an easy or a quick journey, as we shall see in the next chapter. I would describe it as a voyage into the underworld where our greatest fears— namely, the fear of insanity and the fear of death—lie buried. If one has the courage to face those fears, he will return to a new world of brightness from which the clouds of the past have been removed.

CHAPTER 9

FEAR:
THE PARALYZING
EMOTION

All patients in therapy are frightened individuals. Some are not conscious of their fear, others deny it. Very few are in touch with the depth of their fear. In the preceding chapters I pointed out that patients are afraid of their emotions of love, anger and sadness. They are equally if not more afraid of their fear. Even though fear is not a threatening emotion it is a paralyzing emotion. This is especially true when the fear is very big, as in terror. When it is terrified, an organism freezes and cannot move. When the fear is less great it will panic and run, but panic is a hysterical reaction and therefore an ineffective way to deal with the danger. When children are frightened of their parents, who can be irrational and violent, there is no place to run. They become terrified. They are frozen with fear. In the wild when an animal is terrified by a predator and unable to escape, it will generally be killed. Should it escape, the fear will quickly subside and the animal will return to normal. For the child who is frightened of his parents, there is no escape. He must, then, do something to overcome the state of paralysis. He has to deny and suppress the fear. He mobilizes his will against the feeling of fear. He will tighten the jaw muscles in an expression of determination that says, "I will not be afraid." At the same time he will dissociate to some degree from his body and reality and deny that the parents are hostile and threatening. These are survival measures, and while they enable the child to grow up and become free from the possibility of parental attack, they also become a way of life through being structured in the body. The child lives in a state of fear, whether he feels it or not.

While most patients do not sense the degree of their fear, it is not difficult to see it. Every chronically tense muscle is in a state of fear, but the fear is most evident in the tightness of the jaw, in the raised shoulders, in the wide open eyes and in the overall stiffness of the body. One can say of such individuals that they are "scared stiff." When the body manifests an overall lack of vitality in pallid color, muscular flaccidity and dull eyes, the person is "scared to death." To say that the fear is structured in the body doesn't mean that it cannot be released. Releasing the body from its state of fear requires that the person gain consciousness of his fear and the tension and also that some means be found to discharge the tension. In the previous chapter I pointed out that anger is the antidote to fear. The patient has to get mad—that is, angry—but so angry that he feels a little mad or a little out of control. This raises the specter of madness, the fear that "If I let go of control, I will go crazy." Every patient has some fear of going crazy if he lets go of control. In this chapter I will discuss this fear and explain how I deal with it in Bioenergetic Analysis.

One often hears parents yelling or screaming at a child, "You're driving me crazy." That statement indicates that the parent feels at the end of his rope, that he can't take any more of the child's activity, that the stress has become too much. But from my work with patients I have found that the person who is usually driven crazy is the child. I don't doubt that the stress of raising a child in our overactive culture can be overwhelming, especially for parents who are overstressed by their own emotional and marital conflicts. Who doesn't have them? But while stress, if sufficiently strong and continually applied, can produce a mental breakdown in an individual, that situation does not apply to the parent. The reason is that the parent has an outlet for the stress. He can scream at the child and even beat him. The child has no such outlet. He has to stand the abuse, although many have tried to run away. To endure an intolerable stress one has to deaden oneself and dissociate from the body. Children withdraw physically into their rooms and psychologically into their imaginations. This withdrawal splits the unity of the personality and is a schizophrenic reaction.

The split can be a fracture or a complete break, depending on the inherent strength of the child and the severity of the stress. These are quantitative factors which vary with each case. The issue is whether the child can hold together and not break or fall apart. An older child, between three and five, may have developed enough ego strength to resist and not break. Resistance takes the form of rigidity, which allows the individual to retain a feeling of integrity and identity. That rigidity then becomes the individual's psychological level of survival mechanism. The letting go of that rigidity is a very frightening prospect.

Torture in one form or another is used to break a person's spirit, his mind or his body. It doesn't have to be physically damaging. One of the most effective methods of torture is the deprivation of sleep. The mind has no way to recover from the stimulus input which demands a continuous expenditure of energy. Sooner or later the person cracks and the mind splits off from an intolerable reality. Schizoid individuals who become overstressed and unable to sleep will have a nervous breakdown. The triggering factor is a constant stimulation the person cannot reduce or escape. The classic example is a Chinese method of torture in which an individual is buried in the ground with only his head protruding, subject to a continuous drip of water. Eventually the stimulation becomes too much and, since no escape is possible, unbearable and intolerable. At this point the victim starts to scream as a way of discharging the excitation, but if this brings no relief, the individual is driven out of his mind. His control will break and his mind lose its grasp of reality.

A child is more vulnerable than an adult to the torture that could break his mind, if not his body. He has no possibility of escape. Physical abuse is one of the ways in which a child can be broken and we know that it is common. But verbal abuse or emotional abuse is even more common. Many children are subject to constant criticism, which eventually breaks their spirit. Everything they do is wrong, nothing they do meets with approval. The child senses the hostility of the parent, a deep hostility which he cannot avoid nor understand. Esther was a good example of this torture. She was the best-behaved, the politest, the most considerate person I have known. But her life was a disaster. Nothing she did brought her any fulfillment. She failed in her career and in two marriages. Her failure was not due to any lack of trying. She tried to do the right thing or the good thing but it didn't work. It didn't bring her the love she desperately wanted. As a child she had tried to please her mother, to win her love, with no result. Her mother was critical and negating of everything she did.

Esther related an incident that is typical of their relationship. When she was eight years old her mother stood her up and lectured her on her bad behavior. It was not the first time she was redressed for some innocent action, and Esther frowned. This infuriated her mother, who angrily said, "Don't you dare frown when I talk to you." The mother's cold hostility froze the child who was already scared of her. That frozen quality characterized Esther when I first saw her, a mature woman who was depressed by her inability to realize her hopes. The murderous rage against her mother was locked behind a rigid exterior and inaccessible to her. But locking up her anger also cut off her natural aggression, leaving her only the hope that being good would gain her the love she so desperately wanted. It didn't, because love cannot be earned by good behavior. But cutting off her natural aggression she lost her passion.

When her second marriage failed Esther sensed the rage within her. On one occasion she turned on her husband in a wild fury, which made her feel terribly guilty. That anger against him was triggered by his passivity and was a transfer of the anger she had felt against her father, who professed to love her but did not protect her from her mother. In the conflict between mother and child, he took the mother's side. The betrayal by her father almost drove her wild (crazy), but with no support her fury had to be suppressed. Her body was a rigid as a block of wood. As the therapy progressed Esther recognized her problem. She described herself as a "controlled schizophrenic." What she said in effect was that if she didn't hold her feelings under control, she would go mad—angry-mad to the point of losing control and killing someone.

Some persons do lose control and do kill people and themselves. That can happen if the person's ego has become dissociated from the body and its feelings and is too weak to contain the suppressed anger. It is as if these individuals walk about carrying a live grenade of which they are completely unaware. Making them conscious of this suppressed murderous rage reduces the danger of it erupting spontaneously with deadly effect. Accepting one's feelings strengthens the ego and promotes the conscious control of impulses. Accepting a feeling involves more than the intellectual awareness of its existence. One must experience the feeling and make friends with it. When I was a small child a big dog ran up to me and I became very frightened. To help me overcome this fear my mother bought me a stuffed dog which I petted. It helped a little but I never got over my fear of dogs until I lived with them. To help our patients learn to live with their anger in a healthy way is one of the basic aims of therapy.

To one degree or another all patients are controlled schizophrenics. They are all afraid of losing control, of going mad, because as children they were almost driven mad. Gary was a soft-spoken, quiet man whose emotions are subdued. Like Esther he had a very rational mind which governed all his actions, much like a computer. He was, in fact, a computer expert. But there is no joy in a computer or in an individual who functions like one. He had been in psychoanalysis for a number of years but without any improvement in his emotional life. Working with his body to deepen his breathing and getting him to cry made him feel more alive. But he needed to mobilize his aggression and his severely suppressed anger. To do that he had to let go of his unconscious controls. Getting him to kick and scream "I can't stand it" allowed him to break through to a sense of self he had not experienced since infancy.

His story was not complicated. He related, "My mother used to whack me whenever I upset her. She had a short fuse. I recall that when I complained she would hit me. But I couldn't stop complaining, which infuriated her. I couldn't protest. If I cried or said something, she would hit me harder. I remember turning blue in my face, terrified that I would explode. She drove me crazy. I knew she loved me but I couldn't relate to her. She was an unhappy woman and I couldn't make her happy. She drove me crazy."

Gary didn't go crazy. What he did was cut off his feelings by dissociating from his body and going into his head. His defense was different from that of Esther, who became rigid as a means

of control. Gary became more lifeless so that there were no feelings to control. Gary wouldn't kill anybody. He was partly dead
himself. His rage would surface only when he became alive
enough to feel his pain.

And it did surface when, with my support, he felt secure
enough to let himself go a little crazy. Breathing, crying, kicking
and screaming were an integral part of the therapeutic work
almost every session. To gain his voice he had to shut up his
mother's voice, which was now in him telling him what to do,
what she wanted, how to behave, and so on. Twisting the towel
as if it were a neck, he screamed at her, "Shut up. Stop complaining or I will kill you." He also beat the bed with his fists
to smash the image of his hostile mother. Little by little he lost
his fear that if he exploded he would really go crazy. Yes, he
would become mad, but it was the madness of anger, not that
of insanity. In both cases there is a loss of ego control, but in the
former it is through the surrender to the body or the self, whereas
in the latter, the self is also lost.

Any form of overstimulation of a child could lead to insanity
if prolonged enough. One such form is the sexual stimulation of
a child either through physical contact or through seductive behavior. The child has no way to discharge this excitation, which
then acts like a constant irritant in the body. In Chapter 8, I
discussed the case of Lucille, who reported that she was aware
of a constant excitation in her vagina which she could not discharge. As her therapy progressed she became aware that there
was some "craziness" in her personality. She felt confused and
different from other people, which we could trace to her exposure
to a father who was preoccupied with sexuality on one hand
while degrading any expression of sexual feeling or interest on
the other. Her mother behaved like a prude but took secret pleasure in sexual matters. This is a typical double-bind situation in
which two conflicting messages are given to a child: one that

sexuality is exciting, the other that it is bad and dirty. In addition, the parents showed their interest in her sexuality, the father touching her inappropriately on the buttocks. It was enough to drive Lucille almost crazy, but she retained some integrity and sanity through an extreme bodily rigidity. Max, whose case was in a preceding chapter, was almost driven crazy by his mother who, he said, was "all over me." He did not develop the bodily rigidity that characterized Esther or Lucille. Rather than controlling his excitation through rigidity, he acted it out in compulsive sexuality and wild rages. This behavior, however, had no effect in reducing the underlying excitation and the resulting frustration from which he suffered. The frustration stemmed from severe tensions in his body which split the energetic connection between head and body on one hand and between the pelvis and the trunk on the other.

When I look at the bodies of my patients I see their pain in the tensions that bind and restrict them. Their tight mouths, set jaws, raised shoulders, tight necks, inflated chests, sucked-in bellies, immobile pelvises, heavy legs and narrow feet are all signs of the fear of letting go, of a painful state of being. Generally my patients don't complain of pain, although some may experience occasional pain in different parts of their body, such as in the lower back. They do complain of some emotional distress, which is often what brings them to therapy, but in the beginning most assume that it is psychological. Physical pain frightens most people. They react to it as they did when they were very young. They want it to go away. A child's ego cannot deal with pain as an adult's can. If it doesn't go away, they will go away—that is, they will dissociate from the body and withdraw into the head where there is no pain. The point at which withdrawal occurs is when they can't stand the pain in their bodies. By withdrawing from the body they can tolerate the painful situation because it no longer hurts. They have become numb. Normally healthy

adults don't withdraw or split off from the body in painful situations. Their egos are strong enough not to break except in the
most unusual situations, such as while being tortured. When
adults break or split off—that is, dissociate from the body as
Madeline did—it is because the connection between the ego and
the body has been weakened by painful experiences in childhood
or infancy.

Returning to the body is a painful process, but in the reex-
perience of the pain one reconnects to the aliveness and the feelings that one had suppressed in order to survive. No longer a
child, no longer dependent and helpless, one can accept and express these feelings in the security of the therapeutic situation.
But even in this situation patients are too frightened at first to
surrender the ego control that assured their survival.

While the surrender to the body involves the giving up of ego
control over feeling, it does not entail a loss of control over actions
or behavior. This can occur, however, if the feelings are very
strong and the ego too weak. When an individual's conscious
mind is overwhelmed by an excitation which it cannot handle, it
can lose its ability to control behavior. The individual is at the
mercy of feelings which might lead to dangerous and destructive
actions. Such feelings can be a murderous rage or an incestuous
lust. Any person who would act out such impulses is considered
mad or insane and could land in a mental hospital. But the fear
of insanity is more than the fear of committing a heinous act. It
is the fear of the loss of the self. When the conscious mind is
overwhelmed by any feeling, it results in the loss of the boundaries of the self. Like a river that floods its banks, one can no
longer distinguish the river in the mass of water. The river has
lost its identity, which is equally true of the individual who is
flooded with feeling. A loss of identity is one of the signs of
insanity. We are familiar with the fact that an insane person can
imagine himself to be Christ, Napoleon or some other figure. But

the loss of identity need not go to such an extreme. The individ-ual who suffers a nervous breakdown becomes confused about who or where he is or what's going on. It is hard to regard anyone as insane who is aware of his identity and of the reality of time and place. The loss of the self-boundary entails a loss of reality, in effect, a loss of the awareness of one's real self. This, in itself, is a very frightening experience. The person is disori-ented and depersonalized. In the latter condition he is not con-scious of his body, but once the depersonalization occurs, the fear disappears. The dissociation of the mind from the body, which is the split that occurs in schizophrenia, cuts off all perception of feelings. The fear of insanity is tied to the process of dissociation, not to the state of being dissociated—just like the fear of death is really a fear of dying. There is no fear in the state of death. It is the process of giving up ego control that is frightening.

And, yet, it is what we seek in our deepest being since it is the basis for the experience of joy. Many religious rites embody practices which produce an overwhelming state of excitement in an individual, causing him to transcend the boundaries of the self. In a voodoo ceremony which I witnessed in Haiti many years ago, this was achieved through dancing in response to the con-tinuous rhythm of two drums. The young man who danced for nearly two hours to this music ended in a state of trance in which he no longer had full control of his body. I have personally ex-perienced an overwhelming excitement that carried me away to where my sense of reality changed. As a young boy I remember being so excited by the lights, the music and the activity in an amusement park that the scene seemed a fairy world. In a later time I recall laughing so hard at a game that was being played that I couldn't tell if I was asleep or awake. And I have expe-rienced an orgasm of such overwhelming intensity that I felt myself to be out of this world. On none of these occasions was I frightened. What happened could not have happened if I was frightened, and, in fact, they were each extremely pleasurable experiences, joyous to the point of ecstasy.

There is a world of difference between the madness that is passion (the divine passion) and the madness that is insanity. In the former, the excitement is pleasurable, which allows the ego to expand until, at a final moment, it is transcended. But even at this moment, the transcendence is not alien to the ego since it is natural and life-positive. It is a surrender to the deeper life of the self, the life that operates on the unconscious level. Children are not afraid of losing ego control. In fact, they love it. They will spin themselves around until they become dizzy and fall to the ground laughing with pleasure. But if they let go of control in such activities, it is a free act done without pressure. The lack of ego control is natural to a very young child. An infant never had or knew such control; like any animal he functions in terms of feeling rather than conscious thought. As one grows and as his ego develops he becomes a self-conscious individual who thinks about his actions. The imposition of conscious control enables a person to adapt his behavior to larger or more distant goals than the satisfaction of immediate need. But when we act in accordance with our thoughts and conceptions, we are not spontaneous, which eliminates the joy and reduces the pleasure the action could produce. But since this is done in the interest of a greater pleasure in the future, it is a healthy and natural mode of reacting. It becomes a neurotic pattern when the control is unconscious and arbitrary and cannot be surrendered.

Conscious control can be surrendered when appropriate. One cannot give up unconscious control because one is either unaware of the control or unaware of its mechanism and dynamics. This unconscious control affects many individuals who find it very difficult to express their feelings or assert their wishes. They tend to be passive and to do what they are told. Even when they make a deliberate effort to say "no," their voice is weak and the expression lacks conviction. Their self-assertion is handicapped by chronic muscular tensions in the body which constrict the throat, strangling the sound, and by chronic muscular tensions in the chest which restrict breathing, reducing the amount of air passing through the vocal cords. One could say that such an individual is inhibited, that he feels self-conscious about making demands on his own behalf. Generally, the person is aware of his inhibition but he is powerless to release it since he doesn't understand why he is inhibited and doesn't sense the tensions that constitute the inhibition. That problem can be dealt with in therapy, as the following case shows.

Victor, a man about thirty-five years of age, entered into therapy with me because he suffered from a deep sense of frustration about his life. Despite a good mind, a fair degree of competence in his field and hard work, he wasn't successful in his business activities. And this same lack of success characterized his relations with women. Looking at his body I could see severe tensions in his jaw, his shoulders and about his pelvis. The latter tensions denoted that he suffered from a severe castration anxiety. He was aware of how tense he was but he didn't understand the cause and felt helpless to do anything about it. Apart from the above tensions, the most notable feature of his personality was his voice. The sound was soft, subdued and without resonance. It was only a little louder than a whisper. If he tried to shout it would require a strong effort and he would become hoarse. In other respects there was nothing subdued about Victor. He was as intense as he was tense. The tension in his jaw was so severe that he suffered from some ringing of the ears. This tension expressed his determination; everything he did was done with strong determination. Trying so hard, he had little pleasure in his life and no joy. He had to try—he could not let go, he could not surrender.

To understand his problem one had to know his experiences

as a child, for those experiences shaped his personality. Victor was the youngest of three children and, as the youngest, he was the one upon whom the mother focused her feelings. He was her baby and her man and at her service all the time. He recalls that he never could make any demand. In effect, he had no voice in his own life. Unfortunately, Victor's father was also a passive man whose role was to make his wife happy by catering to her. Victor's mother was not a strong woman. She saw herself as a little princess for whom everything had to be done and Victor was elected to serve her. This situation continued until, in the course of his therapy, Victor gained the courage to end it, to assert his independence. He had tried it before but his mother paid no attention to his refusal to serve and he had always capitulated. She simply wouldn't listen to him. One day when she demanded that he drive her to the airport and refused to take "no" as his response, he reached out and put his hand on her throat. It was a spontaneous gesture with no conscious intent to hurt her, but it frightened her so much that she backed off. When he related the incident, I saw the significance of that gesture. Unconsciously, he acted out what had been done to him. He had been strangled as a young child and although it was not physical but psychological, the effect was the same. It was as if a hand had been put around his throat to silence him. As we saw in Chapter 5, one has to reverse the action to free oneself from its effect.

This single action, though a step in the right direction toward freedom and independence, did not resolve his conflicts nor release him from the bonds to his mother. The forces which bound him to her were deep and powerful. They were sexual and he was caught in a web of desire, guilt and rage. Victor was aware of the sexual nuances that underlay his relationship with his mother. She was very seductive with him and totally insensitive to its effect on him. I always inquire early in the therapy about the sexual behavior of all members of the family in my patients' childhood home. In response to my questions, Victor told me how sexually excited he had felt in his contact with his mother. He said, "I couldn't stand the urge and I couldn't contain the excitement. It was driving me mad." But he had to stand it to save his sanity. He put a grip on himself which was still operative when he came to therapy. He set his jaw, held his shoulders rigid and pulled in his belly. But this action didn't remove the charge. It was now locked in his tight, intense body. If Victor had not been able to "stand it" he would have been overwhelmed, which would have flooded his boundaries and destroyed his sense of reality. He would have gone insane. Fortunately, for an adult, the danger is not that great. An adult ego may have a weakness, but it is not the ego of a child—it can now handle a degree of excitement which it could not have handled as a child. Of course, there are limits. Almost any person can be driven insane if enough pressure is put on him to break his ego. On the other hand, a gradually increasing energetic charge could strengthen the ego if the individual has the support of a therapist who would also provide the control which the patient would relinquish.

When the excitement and the tension associated with it becomes too much the body will react spontaneously to discharge it by screaming. The scream is a high-pitched sound which increases in pitch and intensity until it reaches a climax. In the scream the wave of excitation flows upward into the head as opposed to crying, where the wave flows downward into the belly. The sound of crying is low-pitched in contrast to that of the scream. In crying we discharge the pain of loneliness and sadness. It is a cry for contact and understanding. In the scream we discharge the pain of an intense excitation, which may be positive or negative. Children scream with delight when the pleasurable excitement is very great or with fear when there is pain. The scream acts like a safety valve to blow off an excitation which could otherwise "blow the mind" if unreleased. Patients always feel quieter and more open after screaming. And just as we all have something to cry about—namely, the lack of joy in our lives—so we also have reason to scream. For most of us the struggle to survive is too intense, too painful and too tiring, but we stand it because we are afraid to feel the tremendous urge to scream "I can't stand it." We are afraid it will "blow our minds" when in reality it can save them.

On a radio talk show some years ago I described the value of screaming. A listener called in to say that he had been using that release technique regularly while driving home at the end of his working day. He explained that he was a traveling salesman and that by five o'clock he had had it. He felt tense. Screaming in his car while on the road discharged the tension so that by the time he arrived home he felt relaxed and in good spirits. Since then I have heard similar stories from other people. When the car windows are closed, no one can hear the scream. The noise of the car and the traffic drowns all other sounds. I have recommended it to my patients who need to scream but are inhibited by the fear of others hearing them. One could scream into a pillow, but to fully let go, one needs to feel free. My office in New York City is soundproofed.

Many years ago I worked with a woman who felt cut off from life. She explained that she had been married for a short time to a very lovely man who was killed in an airplane crash before her eyes. She had been watching as he came in for a landing in his private plane when it suddenly went out of control and crashed. She must have gone into a state of shock, for she turned and walked away without a cry or a sound. I realized that she had blocked the scream that such an experience should have evoked. While she was lying on the bed, I asked her to try to scream. Only a low-pitched sound came from her frozen throat. To release the scream, I placed two fingers on the anterior scalene muscles at the sides of her neck, which were very tight, constricting her throat. As I applied some pressure to these tight muscles, a scream erupted which she could not control. Her screaming continued after I removed my fingers. Then, when the screaming subsided, she broke into deep sobs which continued for some time. After the crying, she said that she felt that her life was restored to her. Within the year she was married again.

I have used this procedure with very many patients who cannot scream. In almost all cases they respond with screams that are loud and clear. The immediate pressure on these very tight muscles is painful, but the moment the patient screams, the pain disappears because the muscles relax. To scream is so releasing that no patient has ever complained about the procedure, although I always explain in advance what I do and why. My realization of the importance of the scream stemmed from my personal therapeutic experience with Dr. Reich which I described earlier. That scream opened the door to my soul and allowed memories to surface which I had kept buried for decades.

There is another side to the scream that is important to the experience of joy. The flow of excitation in the body is polar, which, as I have pointed out earlier, means that the wave going upward and that going downward are equal in intensity. One aspect of the downward direction is sexual. If one can allow the upward-flowing wave to reach its acme in a full scream, one can also allow the wave flowing down to reach its acme in orgasm. One "blows his top" in a scream and his bottom in the orgasm. Both are powerful discharges. However, the fact that a person can scream once is not a sign of orgastic potency, which depends upon the ability to scream freely and fully all the time.

A scream cannot be forced. If one tries to force the scream, a screech or yell occurs with some scratching of the throat. To scream, one has to let go, something young children can do very easily. Unfortunately, that ability is cut off early in life when a parent can't stand the child's screaming which drives the parent crazy and regards the child as crazy for screaming. Crazy people scream because the internal pressure has increased beyond their ability to contain it, not because they are crazy. They often become agitated for the same reason. Their screaming is a protective measure. If they do not scream to release the pressure they could become violent and kill someone. As a rule the screaming patient is not dangerous. But while the scream is a safety measure, it is not an integrated response to the experience of being driven crazy. Such a response requires the mobilization of the whole body in a meaningful expression. That occurs when the movement of the legs in kicking is coordinated with the scream and with the words, "You are driving me crazy."

Screaming is such a powerful emotional discharge that it was made the basis for two other psychotherapeutic approaches. Best known of these is Primal Scream Therapy developed by Arthur Janov. His book, *The Primal Scream*, caused quite a sensation when it was published, in part because it promised a quick cure for neurosis.[1] The popularity of the book stemmed not only from the promise of a cure but from the fact that it touched a reality in people which had been largely ignored by psychoanalysts and therapists previously. That reality is the existence in all neurotics of deep pain from the early hurts of their lives. Primal therapy is Janov's technique for discharging that pain through screaming, which, temporarily at least, transforms the person into a free individual—one no longer bound by his neurotic fear. Janov saw that the core of neurosis was the suppression of feeling and that this suppression is related to the inhibition of breathing and the development of muscular tension. Reading the book, many people sensed their need to scream to release their pain and re-

[1] Janov, Arthur, *The Primal Scream* (New York: Putnam, 1970).

sponded enthusiastically to Janov's promises of a cure. Patients who "exploded" into a scream after breathing deeply reported feeling "pure," and "cleaned." I had a similar experience in my first session with Reich, and though it opened a window into my deeper self, it was not a cure. I have been on my voyage of self-discovery for fifty years now and while I have found more of myself, I haven't found any cure. Real progress in therapy is a growth process, not a transformation. One becomes a more open, more mature person, but the emphasis is on "more."

Lest there be some misunderstanding, I must explain that expressing feelings is not the whole therapy. Self-discovery requires considerable analytic work that includes the careful analysis of present-day behavior, the transference situation, dreams and all past experiences. Talking together is a major aspect of Bioenergetics Analysis. It prepares the ground for the working-through of a patient's emotional problems but it does not remove these problems on a deep level. I have found from my experience that insight and understanding do not resolve conflicts although they do give the patient the means through ego to deal more effectively with his problems.

No amount of talking or understanding will significantly release the severe muscular tensions that cripple most people. Such tensions block the expression of feeling and can be released only by the full expression of those feelings. But full expression means that the ego must be involved in that expression. Actually, the full expression of feeling not only releases tension but also strengthens the ego and self-possession. One may scream as a child, but when it is done with understanding, one doesn't feel like a child. Adults are grown-ups, which means that they are grown-up children who have all the capacities and sensitivities of the child but also the maturity and self-possession to make their actions effective in the world.

Another scream-based therapy was developed by Daniel Casriel for use with groups.[2] Casriel says, "Screams can release emotions repressed since childhood, and the freedom of release can effect significant positive changes in personality." In addition to screaming, there is the talking about one's life, one's problems, one's hopes, one's dreams. But, as Casriel has learned, the underlying problem in people is "anathematization of basic emotions and encapsulation of feelings behind a defensive shell that is extremely hard to penetrate in traditional psychotherapy situations."[3] I saw a demonstration of this technique by Dr. Casriel at a workshop of group psychotherapists. The participants sat in circle holding hands and each in turn tried to scream "I am angry." Casriel himself participated in this exercise, which seemed to evoke some feeling in the participants. Screaming, as in such exercises, has a cathartic value in releasing some tension, but I don't believe it has much therapeutic value, since the underlying fear of going crazy is not confronted. Such an expression of anger does not involve the whole body and is nowhere near the intense rage which exists deep within the personality of so many people.

The exercise I use is kicking the bed while lying on it, coupled with the scream and the words, "You are driving me crazy." This exercise is a more integrated, total-body expression of feeling. The same exercise can be done with other words, such as "Leave me alone," or "I want to be free." The sound should rise to a full scream. If the patient can give in fully to the exercise, his head will move up and down rhythmically with the movement of his legs and the voice will be loud and clear. When this happens the person will experience a sense of freedom, pleasure and joy from the surrender to a strong feeling. Without consid-

[2] Casriel, Daniel, *A Scream Away from Happiness* (New York: Grossett & Dunlap, 1972) p. 2.
[3] Ibid.

erable practice most patients are incapable of such a full surrender, but each time they do the exercise they gain a further
measure of ego strength. Some patients may get overwhelmed
and feel frightened, but that feeling passes as they quiet down
and sense my support and assurance. It is not an exercise to be
done alone or apart from the therapy situation; its value depends
on the understanding and courage of the therapist to face and
deal with the fear of letting go of control. I have never had a
negative result.

A number of years ago I made a presentation at one of the
local mental health hospitals on Bioenergetic Analysis. As part of
the presentation I was asked to work with one of their patients.
To demonstrate how I work with the body, I gave this patient
a towel to twist while he lay on a mattress and I encouraged him
to express any angry feelings he could evoke. While he was doing
this exercise, I stood at the dais explaining to the audience of
psychiatrists and others the nature of the exercise. The patient
yielded to the exercise with a strong expression of anger both
vocally and in the twisting of the towel. But as he did this, he
"flipped out"; that is, he was out of control. I was watching as I
talked to the audience but I made no move to interfere; however,
the look on the faces of many in the audience was one of shock
at what had happened. I allowed the patient to go through the
exercise, which lasted about five minutes. When it was over, he
recovered his self-possession and I asked him if he was frightened. He said "no," that he was aware I was watching him and
knew what was going on. The experience reduced the patient's
fear of "letting go" to his feelings, which is a necessary element
in the treatment of terrified and schizoid patients. But to work
in this way the therapist must be able to surrender to the body
himself. The patient finds his security base in the competence
and confidence of the therapist.

The kicking exercise is one I use regularly because so many

of my patients who are average people in normal life situations have some fear of going crazy if they give up control and surrender to their feelings. The exercise provides an opportunity for a patient to explore the giving up of control and to gain the ego-strength to surrender to the body and its feelings. Strangely, I have not seen any one of my patients go completely out of control. All of them have been conscious of what they were doing and only allowed themselves to go as far toward surrender as they could handle. But with continued practice, one's ego strengthens to where letting go becomes easier and easier.

I don't believe any amount of rational discussion would significantly help a person lose his fear of going crazy, since the fear is structured in chronic muscular tensions—specifically in the muscles connecting the head to the neck and controlling the movements of the head. One can palpate the tension in these muscles and reduce it somewhat through massage and manipulation, but a significant release that would affect behavior can be achieved only when a person faces his fear and finds that it is unrelated to his present life situation. It was valid when he was a child and his ego was not strong enough to deal with the dangers he faced. But he is not a child now, and if his ego is weak, it is because he is locked into his childhood struggle by his fear, represented by the tension at the base of his head. In the exercise mentioned earlier, that tension is reduced because the head is whipped back and forth by the kicking as one gives in to the exercise and its associated feeling.

Head-banging by children serves a similar purpose. Children who find themselves in a persistent painful situation which they cannot change or avoid and cannot tolerate, bang their heads against a bed or sometimes even against a wall to alleviate the painful tension that builds up in the neck where it joins the head. They are too young to understand why they are forced to take this action and too often their parents are too insensitive to see and understand their plight. But I can understand the intensity of the pressure in a child that would cause him to engage in such a seemingly self-destructive action. They must feel that it is the only way to relieve a pressure that is driving them crazy. I have my patients do the same exercise lying on the bed, using the words, "You're driving me crazy." Since it reduces the tension at the base of the skull, it diminishes the fear of giving up ego-control.

This tension at the base of the head is also responsible for the common tension headaches from which so many people suffer. These headaches develop when a wave of excitation as an impulse of anger moves up the back and is blocked at the base of the skull, causing the tension at the back of the head to intensify and spread over the top of the head like a lid to prevent the impulse from breaking out. As the pressure builds under the lid the person develops a headache. Since the impulse is blocked from expression—that is, suppressed—it never reaches consciousness. The person isn't aware that he is angry and that, in suppressing the impulse of anger, he creates the tension which causes his head to ache. Headaches do not develop when the impulse of anger is very strong since such impulses cannot easily be suppressed. They develop when the power of the suppressive force is stronger than the strength of the impulse. A tension headache will often persist after the impulse has subsided. The muscles relax only very slowly and will continue to ache from the tension. I can often stop such a headache by a gentle massage and manipulation of these muscles by an action which is like unscrewing a tight lid.

But since the fear of one's anger is at the core of the fear of letting go, it is necessary for a patient to face that fear to release it. In effect, I will encourage him to go mad, that is, to become furiously angry. Victor was driven almost mad by his mother's seductive behavior, which tormented him, but his fear of insanity as an adult stemmed from his fear of his murderous rage against her for the loss of his manhood. One of the exercises I use to diminish the patient's fear of his rage was described in Chapter 5 and I will repeat it here in connection with the fear of insanity. The patient sits in a chair facing me and I sit in a chair about three feet away. I have explained to him that this is an exercise in the mobilization of anger. To do that he makes two fists and raises them to me. Then I ask him to extend his lower jaw, show his teeth, and, at the same time, to open his eyes wide, very wide, and to look at me. He is instructed to shake his fists at me, to shake his head slightly and to say, "I could kill you." The most difficult part of this exercise is for the patient to keep his eyes wide open. Often, the wide-open eyes bring up a feeling of fear and the patient closes them. If he feels the fear, he cannot feel the anger. The wide-open eyes have a special effect. They diminish the focus on immediate reality and allow a look of *insanity* to emerge. In almost all cases the patient's face will take on a demonic expression, allowing a feeling of intense anger to come through in the eyes which the patient can then feel and identify with.

The whole exercise takes no more than a minute or two. Once the patient feels his anger, I ask him to drop his fists and relax, but not to let the anger go out of his eyes. If he keeps the anger in his eyes, he integrates this strong feeling of anger into his ego and gains conscious control of it. Having conscious control of it he is no longer afraid to feel the intensity of his anger. Conscious control is manifested in a person's ability to bring an expression of anger into his eyes deliberately. Just as it is possible to express fear in one's eyes by adopting a look of fear—eyes and mouth wide open—so it is possible to express anger by adopting an angry expression. Most people cannot do this at will because they do not have full control of their facial muscles, including those that surround the eye. They lost this natural ability because as children they were afraid to show an angry face to a parent. Although there is a possibility with this exercise that the patient could be overwhelmed by his anger, this can be avoided by an emphasis upon awareness and containment of the feeling. The emphasis is upon the feeling not the action.

A clear expression of anger in the eyes indicates that a strong energetic charge has passed through the body and into the eyes. The flow of excitation in the emotion of anger, as described in an earlier chapter, is upward through the back, over the top of the head and then into the eyes. When I strongly mobilize that expression in my eyes, I can feel my hair stand up along my upper back and on top of my head. One sees the same phenomenon in a dog when it bristles with anger. The importance of this charge for the eyes is that it brings them sharply into focus, improving their vision. As we saw, the opposite movement occurs in fear in which energy withdraws from the eyes. Frightened individuals often feel confused because of a difficulty in focusing. That difficulty disappears through the use of this exercise. However, it must not be expected that doing this or any other bio-energetic exercise once or several times will change a lifetime pattern of fear. The feeling of anger must be integrated into the personality so that its expression is easy, natural and appropriate to the situation. Its expression will then occur spontaneously as the necessity arises. The fact that behavior is under conscious control doesn't negate its spontaneity. We do not think how to walk, eat or write yet we are consciously aware of what we are doing and can consciously control our actions.

One can't have conscious control of behavior if one is afraid to let go of control. This may seem like a contradiction but it isn't. Fear has a paralyzing effect upon the body which, by undermining the spontaneity of an action, makes the action awkward. The conflict between the impulse to withdraw and the impulse to act impairs one's conscious control which supports the fear. There are, of course, historic reasons for the fear. If, as a child, one felt a murderous anger, one would be justified in believing that any expression of that feeling could result in being severely beaten by the parent. In this situation the child has no choice except to inhibit the action and suppress the feeling. But suppressing the feeling fixates the person at the childhood level. The past becomes frozen in the personality but is potentially active. Even in a therapy situation in which all danger is removed, the patient may be still afraid of the consequence that could follow the expression of intense anger.

There is another element in this problem of letting go which also relates to the childhood experience of the individual. Children tend to equate a feeling with action. Wishes and feelings are potent forces. To wish someone dead can be experienced by a child as equivalent to killing the person. Feelings are also seen as enduring. Adults know from experience that feelings change like the weather and even more rapidly. Anger can turn to caring and love to hate, according to the changing circumstances of life. Children, who live fully in the present, do not think in terms of the future and do not, therefore, have a conception of change. A pain is seen as enduring forever. So children often ask "When will it go away?" This kind of thinking extends to feelings. "If I am angry at you," the child thinks, "I will always be angry at you. If I hate you, I will always hate you." Associated with this view is another which equates thoughts with actions: The wish to kill someone is equivalent to the act of killing them. A young child's ego cannot easily distinguish between the thought, the feeling and the action. That distinction becomes possible when the child becomes self-conscious, and his ego recognizes that he has conscious control over behavior.

Analytic therapy is impossible with a young child because he lacks the objectivity required for such a therapeutic process to work. However, many adults also lack objectivity because of their emotional fixation at an infantile level, which undermines the ego and its ability to differentiate clearly between thoughts, feelings and action. An adult could know that though he may have an anger intense enough to kill, he will not act upon it because it would be inappropriate or unwise. The tendency to act out stems from a childish component in the personality. Thus, it is a sign of adulthood when one can have and express the feeling of murderous rage without acting on it or even intending to act upon it. The exercise described earlier, in which the patient sits in a chair facing me and shaking his fists at me while repeating the words, "I could kill you," gives patients an opportunity to experience and develop the conscious control which would allow them to become and act like the adults that they really are.

One other important aspect of this exercise is the relation of the voice to the eyes. Many individuals doing this exercise will shout in a very loud voice the words, "I will kill you," but lack a look of anger in the eyes. Over-accenting the voice diminishes the charge in the eyes. The expression of anger becomes limited to the voice at the expense of the eyes. This is a more childish response since in babyhood and childhood the voice is the dominant mode of expressing feelings. Among adults, however, the eyes become more the dominant mode. Thus, adult anger is most to be feared when the voice is subdued and the eyes are flashing. It is an extension of the philosophy of Theodore Roosevelt, "Speak softly but carry a big stick."

I should emphasize that while the exercises described lessen a person's fear of surrendering to the body, they must be supplemented with other exercises expressing anger. The sensitivity of the therapist to the patient's problem will enable him to choose an appropriate exercise. For example, Victor, whose case was discussed earlier in this chapter, told how his hand went for his mother's neck spontaneously which he recognized as an impulse to strangle her. I can understand that impulse. The tone of a mother's voice as she speaks to her child can be so abrasive that the child can not stand it, or the child can sense a coldness or hostility in the voice that is insupportable. However, generally it is the mother's constant hammering at a child with her voice that can drive the child crazy. In such a situation, if the child cannot escape, its natural impulse is to strangle the mother as the only way to shut her up. Of course, the child cannot act on that impulse and, therefore, has to suppress it. Releasing the impulse in therapy is relatively simple. As mentioned earlier, I give the patients a rolled-up towel which they can twist as hard as they wish. At the same time they are encouraged to voice their feelings. Expressions like, "Shut up, I can't stand your voice, I could strangle you," are appropriate. This exercise gives the patient a sense of power which helps overcome his feelings of being helpless and victimized.

The other strong feeling which is associated with the fear of insanity is sexuality. One can be carried away by an intense sexual passion as much as by an intense anger. A person can be madly in love or mad because his love has been betrayed. But in a healthy individual both feelings are ego syntonic, and experienced as a part of the self. The feelings can be contained, which allows an individual to express these feelings in positive and constructive ways, but containment is only possible when an individual can fully accept his feelings. Acting out, either sexually or in anger, stems from the fear that holding the excitement of the intense feeling would be too much. One can't stand it. One has to do something to discharge the excitement—blow up or get involved sexually, or both. Such behavior is not a sign of passion but of fear—the fear of insanity. That fear is the same as the fear of intimacy. Too much intimacy is frightening because it brings up the specter of being possessed by the other as one was by the seductive parent. What drives a child crazy is the double message: seduction and rejection, love and hate.

To contain a strong feeling is the sign of a passionate nature, whether the feeling is love, anger, sadness or hurt. Containment is the opposite of "standing it." One learns to stand painful or disturbing situations by cutting off feeling. In containment one accepts the feeling and integrates it in his personality. This is not easy to do for an individual whose personality is geared to survival, since survival depended on the suppression of feeling. How does one learn containment when one has been a survivor through most of one's life? In this chapter I described several exercises which help an individual stay with a strong feeling of anger. What can one do about sexual feelings?

The answer may be surprising unless one knows that strong sexual feelings are easier to contain than weak ones. The reason is that a person with strong sexual feelings has more of a sense of self and greater ego-strength with which to contain feeling. Most patients, however, do not fall into this category, which means that much of the therapeutic work is geared to increasing the patient's sexual feeling. This is done by getting the patient to breathe deeply into the pit of the belly where sexual feelings are. Crying deeply is the main mechanism to achieve this. It is also done by helping the patient become more grounded through exercises which mobilize feelings in the legs. All of the exercises previously described help.[4]

[4] See Lowen, Alexander and Rowfreta L. Lowen, *The Way to Vibrant Health* (New York: Harper & Row, 1977; International Institute for Bioenergetic Analysis, 1992) for exercises designed to free the pelvis.

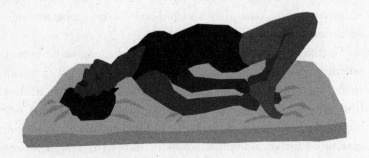

FIGURE 5

Bioenergetic therapists have available a special exercise which increases the charge in the pelvic area without exciting the genitals. I call it the pelvic arch. (See Figure 5.) In this exercise the pelvis is suspended between the head and the feet. To get into the right position you should lie on your back on either a mat or a bed. Bend your knees, keeping your feet 8 to 12 inches apart. Then take hold of your two ankles with your hands and arch up by pulling yourself forward with your hands and letting your head fall back. Only your head, elbows and feet should touch the bed or mat. Push your knees as far forward as they will go and let your pelvis hang heavily.

If your breathing is free and deep, the pelvis will become charged and start to vibrate in an up and down movement. This is a very natural movement in which individuals who are relatively free from tension in lower back, buttocks and thighs experience very pleasurable and exciting sensation in the pelvis which, though sexual, are not genital. But many persons become frightened when the pelvis develops this involuntary movement. If it becomes strong the person may feel out of control, which can be scary. The exercise can also be painful if the muscles along the front of the thigh (quadriceps) are contacted and tense. When that happens one can stay with the exercise if it is not too painful or return to the starting position and try again. Doing this exercise regularly stretches and relaxes the tense muscles, allowing more sexual feeling and pleasure to develop. One can understand this exercise if one knows that this culture is based on two injunctions—don't lose your head and don't let your ass run away with you. Or, don't be carried away by your passion. But unless one can be carried away by an overwhelming feeling one cannot know the key to life. People who have discovered this key have an inner morality that is rare in this culture.

It is very important to remove the patient's guilt about sexual feeling, which is the core of the analytic process. As a patient's sexual feelings become strong they will show in his or her eyes, since both ends of the body become more alive and excited. Bright eyes are the sign of strong sexual feeling. Now the patient can practice holding that charge in the eyes as they make eye contact with the therapist and with other people with whom they interact. This is not easy since most people are ashamed and afraid to reveal their sexual feelings. It is especially true of individuals who have been sexually abused. It is most important for the therapist to accept the patient's sexual feelings without becoming involved, since this would violate the therapeutic relationship.

The next step is logical. The patient must be encouraged not to engage in sexual relations unless there is a strong feeling of love between the two individuals. Consciously holding back when one has sexual feeling promotes containment and increases ego-strength. If the feeling is strong, masturbation provides a proper outlet. Conscious control of strong feeling is the mark of a mature individual who has self-possession or has gained it through the therapy. Such an individual has no fear that the expression of a strong feeling would make him look or feel crazy.

To fully understand the fear of insanity that exists in so many people we need to be aware of how our culture plays a role in driving people crazy. We live in a hyperactive culture which overexcites and overstimulates everyone exposed to it. There is too much movement, too much noise and sound, too many things and too much dirt. A recent cover for *New York* magazine showed a distraught man holding his ears and screaming, "The noise is driving me crazy." One can survive without literally going insane, but to do that one has to deaden the senses so that one doesn't hear the noise or see the dirt or sense the continuous movement. But a similar hyperactivity goes on in today's home with its television sets and appliances. In this culture one can't slow down or quiet down. The hyperactivity is fueled by the same frustration which drives the hyperactive child—namely, the inability to stay in contact with the deep, inner core of one's being, the soul or spirit. Our culture is outer-directed in that we are trying to find the meaning of life in sensation, not in feeling; in doing, not being; in owning things, not one's self. It is crazy and it makes us crazy because it cuts us off from our roots in nature, from the ground we stand on, from reality.

But I believe the worst element in this culture is the over-focus on and exploitation of sexuality. We are continually exposed to sexual images which are exciting but also frustrating since there is no possibility of immediate discharge. This sexual over-stimulation forces the individual to cut off sexual feeling in order not to be overwhelmed or to go out of control. But since feeling is the life of the body, the neurotic individual whose sexual feelings have been suppressed is driven to act out sexually in search of excitement and feeling. This generally takes the form of rape, the abuse of children or pornography. We cannot deal with this problem by moral lecturing or teaching since it stems from a loss of contact with nature and with our own true nature—the life of the body.

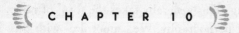

CHAPTER 10

THE FEAR
OF DEATH

Every patient is consciously or unconsciously afraid of letting go of ego-control and surrendering to the body, to the self, to life. That fear has two aspects: one is the fear of insanity, the second is a fear of death. In Chapter 7 we saw that the fear of insanity stems from a subliminal awareness that too much feeling could overwhelm the ego and result in madness. That awareness is connected to the experience in childhood of being driven almost crazy by the hostility, the harassment, the confusion and double messages to which so many children are subjected. Similarly, the fear of death is connected with a very early experience in which the child senses that it faces death, that it could die. That experience is so shocking to the organism that it freezes in terror. Death does not occur, the child recovers, but the bodily memory cannot be erased, although it is repressed from consciousness in the interest of survival. The body memory persists in the form of tension, alarm and fear in the tissues and organs of the body, especially in the musculature.

Looking at a person's body, one can evaluate that fear. If the body is very rigid, one can describe the person as being "scared stiff." This is not just a metaphor, it is the literal expression of the body. If the rigidity or tension is coupled with a lack of vitality in the body, one could say that the person is "scared to death." One could be scared "out of his wits," which is the schizophrenic state. In other individuals, the tension is most apparent in the area of the chest, which is over-inflated, denoting an underlying panic.[1] Most individuals do not sense how frightened they are unless they are threatened by a loss of love or security. But the fear is always there beneath the surface, inhibiting their surrender to life and to the body. They are survivors who walk a narrow path between too much feeling, with a fear of insanity, and too little feeling, with a fear of death. I have found this fear of death in all the patients I work with as a deep and unconscious resistance to deep breathing and surrender.

I first encountered this fear of death in a participant in a workshop for professionals interested in Bioenergetic Analysis. This participant was a man about thirty-five years of age. He was lying over the bioenergetic stool during a break in the sessions and as I passed by I saw in his face a look of death which I immediately thought could only have come from some very early encounter with death. When the sessions resumed, I asked him if he could discuss my observation in front of the group. He agreed. The story he related is that when he was about a year old, he almost died. The circumstances as he learned them from his parents were that he had stopped eating and the weight loss had become so severe he had been taken to a hospital in a critical condition. Questioning revealed that he had been breastfed, which was stopped just before he became ill. He had no knowledge of any connection between these two events but I was strongly convinced that the loss of the breast was the loss of his

[1] See Lowen, A., *Love, Sex and Your Heart.*

world and that he would not accept a substitute. Not every child who is weaned from the breast goes through a reaction so severe it can threaten his life, but weaning can be very traumatic to a child, as many mothers who have breastfed their children know. Much depends on the sensitivity of the parent to the child's distress.

Over the years I have learned from many of my patients that as children they were afraid of death, afraid that they would die. These fears generally arose at night when they were alone in their room or bed. I recall as a preadolescent being afraid to fall asleep, out of fear that I might die during my sleep. Consciousness was my safeguard that I would not die; it was my way of keeping control. Why should any child have this thought? Where did it come from? Had I ever experienced a life-threatening illness or condition? I knew that I had had the usual childhood illnesses, but so much of my earliest years was buried and forgotten by the kind of repression which affects most of us. Even though I had some experiences of joy, there was a sadness in me which my earliest pictures reveal. I did not have a happy childhood. And I believe this is true of most people.

Children, especially infants, need unconditional love if they are to grow up into healthy adults. In fact their very survival depends on a loving connection with an adult. Infants in a nursery who are fed and cleaned but not held or played with can develop anaclitic depression and die. Pleasurable, physical contact excites the child's body, stimulating all functions, especially respiration. Without that contact the basic protoplasmic activity of expansion and contraction, as in respiration, slowly decreases, leading to death. The infant has that contact in the womb, and if it is not reestablished after birth, the newborn organism goes into shock. Of course no one believes that a newborn human infant could survive without care, but we do not appreciate how dependent every child is on a loving connection to a mother-figure. Any break in that connection, or even the threat of a break, results in a shock to the organism. Shock has a paralyzing effect upon the body's basic functioning, which can be fatal if the state of shock is deep and prolonged. But every shock is a threat to the living process. A sudden loud noise can cause a baby to go into shock momentarily. Its body will stiffen and it will stop breathing. This reaction, known as the startle reflex, is present almost from birth. As the shock passes the baby begins to cry, which restores its breathing. Of course as the baby grows the organism becomes stronger, and it is not shocked as easily by sound. But even adults can be startled by a sudden loud noise and momentarily go into a state of shock.

Every time a parent yells or screams at a young child it must have a strong negative effect on his body. One can tell when the child suffers a shock because his body stiffens and then breaks down into sobs. If he is yelled at often enough he will not react, because he has adapted to the stress. He does this by maintaining the state of rigidity or tension. He can no longer be shocked because he is in a continual state of shock, which we know because his breathing is no longer free and easy. In this case the shock is not just from the sound but from the threat to the child's loving connection with its mother. An angry or hostile look, a cold manner or the statement "Mother doesn't love you anymore" could have this same effect. Physically hurting a child by slaps, blows and spankings traumatically shocks the organism because a young child's ego has not developed to where it can understand that a hurtful action by a parent is not a final denial of love. He reacts to the action as if it were a threat to its life. The number of shocks that the average child in our culture suffers is great, and in some cases children succumb to destructive treatment by their parents.

Some parents are hateful, but most swing between love and hostility. An angry outburst would be followed by some expression of love which reassures the child and restores its hope that the loving connection to the parent is secure. As the child survives and grows older he will do whatever is necessary to maintain a connection, even if it involves the surrender of the self. But a connection based on submission is never secure, since the child will try to rebel and the parent will keep the threat alive. No parent fully trusts a submissive child since he or she knows that submission covers hate. And the child knows within his being that he is hated. For the child, survival demands the denial of this reality—denial of the threat to his life, the fear of death and his feeling of vulnerability. Survival also demands that he make an effort to maintain the vitally needed connection with the parent. That becomes the big struggle which will engage the individual as child or adult for all his life, since this pattern of behavior now becomes structured in the personality and in the body as an habitual attitude.

What we see structured in the body is the state of shock manifested in the inhibition of breathing. On the surface the individual doesn't seem to be in shock. To most eyes he appears to be functioning normally. His breathing seems to be regular and without difficulty. But this is because the breathing, the life, is shallow and on the surface. The state of shock exists on a deeper level in the repressed unconscious, in the loss of passion, in the fear of surrender and in the tension and rigidity of the body.

The surrender to the body involves nothing more than allowing its full and free respiration to occur. The fear of surrender is connected to holding the breath. One can block free respiration by restricting inspiration or expiration: in the former, one holds against taking in air; in the latter, against letting it fully out. Both act to limit the amount of oxygen the body absorbs, which reduces metabolic activity, decreases energy and diminishes feeling. The restriction of inspiration is found in schizophrenic or schizoid personalities, where it is associated with an underlying terror, the effect of which is to paralyze all action. In contrast, the neurotic individual has difficulty in letting the air fully out. The fear which blocks full expiration is panic, which differs from terror in that the panicked individual seeks to escape from the danger, whereas the terrified individual is frozen. The danger is different in the two cases. In terror the danger is experienced as an overwhelming threat to one's existence, whereas in panic the danger poses a possible threat to one's existence. For a young child panic results mostly from the loss of the connection to the mother or parent. Thus, a young child who is separated from his mother either by a crowd or by being left with a stranger would panic, scream or cry loudly in the attempt to restore his lifeline to his base of security. When the connection is reestablished, he will hold on to the security for dear life. He also holds on to his air, that is, he maintains his chest in the inflated position. This position, which allows only a shallow breathing, suppresses the feeling of panic by providing a false sense of security stemming from the ability to hold on. In the neurotic character structure, the fear is suppressed and the individual is generally unaware of the degree of his fear. In the schizoid structure, the attempt to suppress the fear is less effective because of the weakness of the ego, and the individual is often aware of his fear. However, in both types it is manifested in the body—in the schizoid body by a collapsed chest and in the neurotic by an inflated chest. The collapsed chest is soft, the inflated chest is hard.

These distinctions are important for an understanding of the fears that prevent the surrender to the body. Terror inhibits any aggressive action, and since breathing in is an aggressive act in which the organism sucks in air, the strength of that action of sucking air is a good measure of the organism's ability to be aggressive, that is, to reach out for what it needs and wants. Breathing out, on the other hand, is a passive action, a letting go, a relaxation of the muscular contractions that inflate the chest. Because of his fear of letting go, the neurotic holds on in adult life to people as in childhood he held on to his mother. All very young children hold on to their mothers, to her body or to her clothes, since she is their basic security. As they grow older and stronger the urge to be independent and to stand on one's own feet becomes dominant. The security represented by the mother is replaced with a sense of security in one's self and in one's body. But the security in the self develops only to the degree that the child feels secure in his connection to his mother. Every time that connection is threatened, his body contracts and his breathing is impaired. The feeling of need for her is reactivated and his dependence on her increases.

Panic can develop whenever one's life is threatened. In panic there is a loss of control of one's actions, as the person runs about wildly to escape the threat, breathing quickly and shallowly. Some people panic more than others when faced with a threat to life, while a few individuals with a strong sense of inner security can maintain ego control in such situations and not panic. On the other hand, there are people who panic in situations that are not life-threatening, like traversing a high bridge in an automobile, or finding themselves alone in a crowd. Panic disorder is a recognized neurotic condition and includes individuals who cannot leave their home alone without experiencing intense panic. If we wish to understand why a person would panic when finding himself alone away from home, we must recognize that the feeling is one of being in a life-threatening situation. Since the feeling is irrational, we must assume that the situation evokes a body memory of when one was in such a life-threatening situation as a child. Perhaps the most common situation is a mother's negative response to a child's crying. When a child cries, he is calling to his mother out of need. Her failure to respond, whatever the reason, is experienced by a child as a loss of his mother, which is life-threatening. In his desperation he will cry harder and harder, louder and louder, longer and longer, driven by need. Such crying uses up the child's energy and he may suddenly find himself in a state of panic, unable to breathe easily and gasping for air. To safeguard his life, the body cuts off the crying by holding the breath in to gain control. As he does so, the feeling of impending death disappears temporarily. The child falls asleep exhausted and as time goes on the memory of this experience will be repressed, but the body does not forget.

One experience does not lead to neurosis. Unfortunately, it is true that many children in our culture not only suffer from a lack of the nurturing and support that would enable them to grow into independent and mature adults, but also are threatened by their parents with punishment for innocent acts. Most parents themselves grew up in homes in which there was violent behavior by one or both parents. Lacking their own inner security and stability, many parents act out their frustration and anger on their children, who live under the constant threat of the loss of love and in a constant state of fear. This fear is manifested in the many emotional or physical disturbances children suffer. It is not surprising that the struggle for survival is all they know.

One might argue that the cases I see are unusual and do not represent the average childhood. But no one, except those living in it, knows what the average home is like, and even those in an extremely unhappy family may deny the degree of unhappiness which they experienced. The people who do come to me as patients are average people whom no one would regard as insane or otherwise severely disturbed. They work, they may be married, have children and are relatively financially well off. But when one gets to know them one becomes aware of a degree of struggle and unhappiness that is shocking. The following is the story of one patient's childhood and life.

Alice is thirty-two years old and has been married for ten

years. She related the following: "As a child I was always scared
and nervous. I felt hated by my mother and rejected by both
parents. My mother criticized me constantly. I felt very alone,
worthless and depressed. Any show of emotion or problem was
turned against me by my family as my fault and then ignored.
I felt I wasn't good enough and would never measure up.

"As a teenager I tried to be perfect but I developed insomnia
and stomach problems. I became anxious and depressed. I was
given Halcyon for my sleep problems and medication for my
stomach disorder. I have been in several therapies over the years,
both individual and group, and I have made some progress, but
I still need my sleeping pills to get the rest so I can function and
carry on my life. I still suffer from constipation, muscular tension
in the area of my diaphragm and I have a growing sense of
loneliness and emptiness that makes me feel isolated both in my
marriage and in my life."

Can one doubt that the experiences of her childhood were
responsible for her problems as an adult? Alice didn't doubt it.
But with all the insight that she had gained through her therapy,
she still felt, at the time I saw her, that she was not able to get
better and free herself from the past. This posed the question—
what fear tied her so firmly to her past that despite her best
efforts she could not free herself to live fully in the present? But
before one answers that question, one needs to understand the
present more fully.

When she came to me, there was no joy in Alice's life and
very little pleasure. She suffered from a severe anxiety that she
would fail, which seemed to have some basis in the fact that in
the past ten years she lost many jobs because of her inability to
function. But at the same time it was clear that given the degree
of her anxiety it was almost impossible for her to function well.
She was caught in a vicious circle. Anxiety made it almost impossible for her to hold a job, which in turn increased her anxiety.

Caught in this trap, Alice's life was a desperate struggle to survive.

The clue to the resolution of her conflict lay in her statement that in her adolescent years she had tried to be perfect. That effort had failed—as it must since no one can be perfect. But if she didn't try to be perfect, she felt worthless and hopeless. It was hell, and I could understand her desperation to get free. But how? Helping her become stronger so she could try harder to be perfect would only lead to more failure and a greater desperation. Any effort, any trying, was doomed to failure. To give up trying to change, to accept herself, was frightening, but it was the only way to some sanity.

What Alice had to accept at first was the unhappiness that could not be denied and her need to cry. She had much to cry about. When I pointed this out to her, she remarked that she *had* cried a lot. This is a common answer and undoubtedly true but the question is—how deep was the crying? If the crying is as deep as the pain and sadness, it will fully release the person. The pain in Alice was deep in her belly, associated with her intestinal distress, but it was also felt in the area of her diaphragm where it was due to a band of tension which blocked both her breathing and her crying from going deep into the belly. That region is where our deepest feelings are: our deepest sadness, our greatest fear and our most profound joy. The sweet melting sensations that accompany true sexual love are experienced deep in the belly as a glow which can extend through the whole body. Pleasurable sensations in the belly are experienced by children on the swings and seesaws that they enjoy so much. But just as it is the locus of joy, the belly is also the place where the sadness of despair is felt when there is no joy.

To reach that joy Alice had to open herself to her despair. If she could cry from that deep despair, she would touch the joy that gives life its true meaning. While we must recognize that despair is frightening, we should also know that it stems from the past and not from the present. Alice was in despair because she could not be perfect and win the approval and love of her parents. Her despair continued into the present because she was still struggling to overcome what she regarded as her faults and weaknesses in order to win that love. In effect she was trying to "overcome" her despair, which could not be done since despair is her true feeling. One can deny the despair and live by an illusion, but that will inevitably collapse and drop the individual into a depression.[2] One can try to rise above the despair, which undermines one's sense of security; or one can accept and understand it, which releases the person from fear.

Accepting the despair means feeling it and expressing that feeling in sobs and words. Crying is the body's statement; words come from the mind. When appropriately matched, they promote the integration of body and mind which releases guilt and promotes freedom. The right words are important. "It's no use" is a key phrase. "It's no use to try; I'll never win your love" is the statement which expresses the understanding that the despair is the result of a past experience. Most patients, however, project their despair into the future. When they first feel the despair it is often expressed as "I'll never have anyone to love me" or "I'll never find a mate". They do not understand that one cannot find love no matter how hard one looks, and that the more desperate one is, the less likely it is that another person can respond with positive feelings. True love is the excitement one feels in anticipation of the pleasure and joy one would have from closeness and contact with another. We love people who make us feel good; we avoid those whose presence is painful.

---

[2] See Lowen, Alexander, *Depression and the Body* (New York: Coward, McCann & Geoghegan, Inc., 1972; Arkana, 1993) for an analysis of the role that illusion plays in the genesis of depression.

Alice's problem was that she was afraid of her despair, because on a deep level it was linked with death. She had lived almost all of her life on the edge of despair, too frightened to feel it. She was like a person at the seashore who only wets her feet out of fear of being overwhelmed by the force of the sea. The sea is a symbol for our deepest feelings—sadness, joy, sexuality. It is the source of life and only by surrendering to it can one live life fully. Going deep into one's despair is to plunge into the depth of the belly, which, representing the sea, is also the source of life. No adult ever drowned in his tears, although a fear of drowning underlies panic. An infant who is cut off from any loving contact will die, a very young child in this situation could die because his body needs the contact and support of a mother figure. The child who comes close to death because of insufficient loving support and survives becomes a neurotic who will live on the edge of despair and panic throughout life, unless he is released from fear by reexperiencing the early trauma and discovering that he will not die.

It must be realized that talking about the fear of death, while necessary to help the patient understand his problem, is not enough to remove the fear. Telling a child not to be afraid of the dark because there is no one there does not help much because, while there is really no frightening figure in the darkness of the room, there is such a figure in the darkness of the child's unconscious. To go into one's own unconscious is to descend feelingwise into the belly through deep breathing. As the respiratory wave of expiration flows downward into the pelvis one can sense the feelings which are locked in this area. One could feel that one wasn't loved and that one could have died, but, sad as that awareness is, one could also realize that one *didn't* die. For an adult not to be loved is not a sentence of death. One can love and surrender to one's self. The mature woman of fifty, mother of several children, who told me "If no one loves me, I will die" is a pathetic individual who is afraid to live as much as she is afraid to die.

At the time I worked with Alice I did not understand how deep the fear of death is, so while I could help her with the neurotic aspect of her problem—namely, the need to be perfect —I was not able to help her face the underlying fear of death which motivated this unrealistic drive. She made some progress in her therapy with me in terms of understanding her issues and feeling a little stronger in herself, but I was not able to help her break through to her deepest self. Most therapists would consider this outcome satisfactory, but without a solid base in one's self and one's body there is the danger of a relapse into despair, since one cannot feel the joy of life. This is not to say that one can consistently achieve the breakthroughs that would release a patient from his fear of death and his despair. But I do believe that it is very important for a therapist to understand the depth of the despair in the modern individual, and to have available the means and understanding to deal with it. The following case illustrates the principle I use to deal with this problem.

Every patient needs to break through the barrier that the fear of death creates, and Nancy achieved this breakthrough. She was a fifty-year-old woman who had a borderline personality. Throughout her life she had been anorexic and her functioning could only be described as marginal. She achieved this breakthrough after several years of therapy during which she gained the will to fight for her life. We had done considerable work to help her breathe better, express feelings of sadness and protest and stand up for herself in a negative life situation. But her feelings never became strong enough because her breathing never went deep enough.

As she lay over the stool, breathing and then making a sustained sound, she stopped the sound just where it could have broken into deep sobbing. She felt very frightened and said, "It's getting very dark. I feel I am going to die." Such a feeling could frighten anyone, but why did she feel that she was going to die just when her breathing was actually stronger? The answer is that her deeper breathing had touched a fear of death that was always in her. Nancy had almost died as a child. The story she told me is interesting: When she was about two years old, she was a plump, good-looking little girl. Her mother, seeing her gain weight, became frightened that she would grow into a fat lady, as her own sister had done. Acting from this fear, the mother hounded the little girl about eating and so terrified her that the child lost her appetite and could not eat. As the child lost weight and became thin the mother panicked and urged her to eat, but to no avail. The little girl ended up in a hospital in critical condition.

I was sure that her anorexia resulted from this experience. She was still terrified of gaining weight when she started her therapy with me. But gaining weight was very difficult for her. For one thing, it meant she would have a body and be somebody. That might result in a confrontation with her mother, of whom she was terrified. An important breakthrough happened several sessions after the one in which she experienced her fear of death. I had reassured her that there was no danger she would die. What happened, I explained, was that as she breathed deeper, she felt her terror and stopped breathing, cutting off the flow of oxygen to her brain. This produced the sensation of darkness. The only thing that could happen as a result is that she might pass out, in which case her breathing would resume spontaneously and she would recover full consciousness. When we returned to this exercise in the next session, she still experienced the darkness and fear of dying, but to a lesser degree. We had established a strong therapeutic alliance at this point which enabled her to trust my leadership. On the third occasion, as she was lying on the bed kicking and trying to scream, a strong sound broke through and she went into deep sobbing which came from her belly. When her crying ended, she exclaimed, "I didn't die! I didn't die!" She felt that she had gotten through a fear which had haunted her, binding and constricting her life. Her courage in dealing with her life situation increased considerably, for she had gained some feeling in her belly—which we could describe as "having guts." But this is not to say that all the fear has left her. She had faced her fear of death, she had entered the underworld, and now she had to work her way through it.

One of my patients related an incident that happened to his daughter, age five. She was playing a game of ball with her parents and enjoying it immensely. Her younger brother, age two, who was watching, wanted to take her place. She refused to give him the ball, and when the parents insisted, she threw the ball at him. It didn't hit him, but her father reprimanded her severely, saying she shouldn't do that because she could hurt him. The reprimand was like a shock and she began to scream. Her father, thinking her reaction was irrational, told her to stop screaming, which only made her scream louder. Wanting to teach her a lesson, he put her into a large closet and shut the door, saying that she could come out when she stopped screaming. After some minutes she did stop but she didn't come out. Alarmed, her father opened the door and found her on the floor, white and unable to breathe. They rushed her to the hospital where the doctor administered a bronchodilator. She had suffered an asthmatic attack which could have caused her death. Unable to stop her screaming and fearful that she might never get out of the closet, she went into a panic reaction in which her bronchi contracted, making it almost impossible for her to breathe. The little girl was in a state of shock.

I have worked with many asthmatics. As they do any of the exercises which deepen their breathing, such as crying, kicking, or screaming, they begin to wheeze and immediately bring out their inhaler, which alleviates the bronchial spasm temporarily, enabling them to breathe more easily. It doesn't, however, remove the tendency to spasm, which is a panic reaction when their breathing gets deeper. Since they become very frightened at the wheezing that marks the beginning of an asthmatic attack, they attribute their fear to the inability to breathe. This is partly true, but it is equally true that it is the fear which creates the inability to breathe. It is the fear of being rejected or abandoned for crying, screaming or being too demanding. This vocal expression which has been suppressed in the interest of survival is activated by the deeper breathing. Once they understand this dynamic, their fear diminishes. I can then encourage them to give in to louder crying and screaming, which they can do without becoming asthmatic. Even when some wheezing develops I discourage the use of the inhaler, assuring them that if they don't panic, they will be able to breathe easily. To their amazement it works in nearly all cases.

Alice, whose case I described above, was not typical of a panic patient. Her chest was not inflated and she had more difficulty breathing in than she had breathing out. Her fear was deeper and verged on terror, which is the response to the mother's hostility rather than to rejection and abandonment. Alice could be described as a borderline patient with a strong tendency to schizoid splitting and dissociation. Her underlying fear was of being killed rather than rejected or abandoned. It is a deeper, more intense fear and requires a greater mobilization of anger to overcome it.

In panic there is also a fear of death, but to a lesser degree. To help patients get in touch with their panic I use the technique described in Chapter 3: The patients lie over the bioenergetic stool and make a sound which is sustained as long as possible. At the end of the sound, they try to sob. When they break through to the sob, they encounter their fear of drowning in their sadness or of being overwhelmed by their despair. To defend against these feelings, the body attempts to inhibit its breathing. The chest wall becomes tighter and the bronchia contract. At this point the patient feels the panic. Lisa, who experienced this panic, remarked, "I feel that I am not able to breathe. My chest and throat feel very tight." But she did not recognize that she was reexperiencing the trauma of her childhood. She added, "I know that feeling (the tightness in her chest). It's a hurt so deep that I don't know if I want to die or that I will die. It's such a silent pain, a private hell." She then explained that she was left alone as a child. Neither parent was interested in or even aware of her struggle and unhappiness. They wanted a happy child and Lisa put on a happy, smiling mask to hide her sadness and despair. Letting herself cry deeply, she felt a sense of freedom in giving up the mask. Lisa had never married and had not experienced the ecstasy of love. She dared not open her heart to love; there was too much pain in it. But only when that pain is experienced does the fear of it leave. At the time she expressed the above insights, she had met someone whom she felt she truly loved.

Sally was a woman whose body between her head and her pelvis was so narrow and tight that I saw her as a lady in a straitjacket. She had a well-formed, strong head, and a broad, good-looking face. Her legs and feet were shapely and strong. In view of her broad face and healthy legs, the narrowness of her trunk could not be regarded as a developmental failure but could only be seen as the result of traumatic experiences in childhood which operated to constrict her chest and pelvis. This constriction was so strong that her breathing was severely restricted. Despite this reduction in respiratory volume Sally was not weak. Her musculature was well-developed and able to sustain a strong effort. The tension in her trunk served a special function, namely to restrain any wild or violent outbreak. Straitjackets are used in mental institutions for that purpose. Sally was a lady in a strait jacket.

Sally entered therapy to learn how to cope with the threatened breakup of her marriage. The marriage was not a happy one, yet the prospect of being alone frightened her. She described her husband as unreliable. He was constantly changing jobs, and Sally suspected that he was unfaithful. He was more a boy than a man. Sally was the responsible one in the family, earning money, running the house and taking care of the child. The marriage couldn't work well because Sally felt used and her husband felt trapped. He responded to Sally's demands for a more responsible attitude with promises which he did not keep. When they finally separated, Sally became depressed with suicidal thoughts. She could not see herself living alone, and she could not envision the possibility of finding another man. Despite the fact that men were attracted to her, she felt desolate. On a deep level she saw herself as an abandoned child. On the surface she continued to work effectively and to handle her life situation.

Coping, however, is not what therapy is about. Life has to be more than a matter of survival. We need to find some joy in life, otherwise we will fall into a depression which can make even survival problematic. Sally felt no joy; the severe tightness of her body precluded any feeling of freedom and ease. She had to be released from her muscular straitjacket, but to help her do that one had to know the events in her life that had led to her being in a psychological straitjacket and understand the forces in her personality that kept her so bound.

When I questioned Sally about her history, she related a story her mother had told her. Sally was the last of three children, eight years younger than her next older sister. At her birth, which took place at home in a rural village, she looked so weak and blue that her mother believed she would die. She was, therefore, laid aside. But she didn't die, and she actually turned out to be a vital child. Sally always attributed her fear of abandonment to this incident, but as one got deeper into her history, other aspects of this fear of abandonment emerged. When she was four years old, the critical oedipal period, her father left the house. Her mother had accused him of being irresponsible and having other women. In this regard Sally's own experience seemed to replicate her mother's. However, Sally's father did visit the family from time to time. Sally recalled how excited and happy she was to see him and how crushed she felt when he would leave again. She expressed this theme often during the therapy. At one session she said, "The minute a man leaves me, I feel I'm going to die. When I have a fight with a man, I feel that if he leaves, I'll die."

When Sally got into her crying, she would say, "Don't leave me, Daddy." She recognized that she had looked to her father for love, support and protection, which she felt she didn't get from her mother. When he left the family, her mother had to go out to work and Sally was left with her grandmother, of whom she was terrified. She had a dream that she was standing on the shore of the sea and saw her grandmother coming toward her. She felt that she would be killed. In the dream she had an impulse to walk into the sea and drown herself. Sally also remembered an incident in which her grandmother washed her hair in very hot water, which caused her to scream in pain and try to lift her head out of the water. But the grandmother pushed her head back down into the water, saying it had to be hot enough to kill the lice. Her grandmother was very severe with her and threatened to kill herself if Sally was not a "good," obedient girl. To make the threat effective she always carried on her person a small bag supposedly containing poisonous herbs, which she threatened to eat whenever Sally cried or protested. The inability to cry or protest strongly against mistreatment was still present in Sally's personality due to the severe restriction in her breathing caused by the tension in her chest and throat.

To free Sally's body from this tension was no easy task, since the severe tension immobilized her aggression. To rebel was to invite disaster which, in her feeling, was to be abandoned or to be killed. She understood well enough that her problems stemmed from early childhood and that her fear of being killed was groundless as far as her present life was concerned, but her fear of abandonment seemed to have a reality in the present. Most patients panic at the thought of being alone, of not being loved, despite the fact that many have lived alone for part of their lives. Sally countered this fear with the hope that she will find someone who will love her and to whom she will be submissive as she was to her grandmother. Submissiveness, however, undermines the relationship and revives the fear of abandonment. If the other person has the same fear and the same need for contact, it becomes a co-dependent relationship, which is only a substitute for love. Sally and her husband were in such a relationship.

Following the break-up of her marriage Sally fell in love with another man, who turned out to be just like her husband and her father—irresponsible and dishonest. He came on very strongly, proclaiming a love for her that turned out to be more words than feeling. When the relationship broke up after Sally discovered that he lied, she went into a deep despair, feeling that she couldn't go on, that she was going to die. It helped when I pointed out to her the person who is looking to be saved ends up being damned. She didn't *need* a man and she was very capable of standing on her own strong feet, but she resisted taking that position because it required facing her despair—the despair that her father would never return and that she would never find a man to love her. She could see the reality of her situation consciously but not emotionally, because she felt it as too painful and too frightening. She was afraid that accepting that she had been betrayed by her father, her husband, and this man, would unleash a murderous rage against them which would explode in a fury of anger that she would experience as madness. To prevent that from happening she had locked herself into a psychological straitjacket. But while releasing her rage might have been dangerous when Sally was a child, it was not true for the woman Sally was now nor the situation she was in. As my patient, she could release her rage by kicking and beating the bed for all she was worth. This she did, and she also expressed an intense anger against me for not taking her feeling of despair as seriously as she wished. She resented my statement that she had within herself the means to work through her problem. In this feeling she identified me with her grandmother, who had demanded a mature attitude from a very sad and frightened child. I, too, was pointing out her need to be more realistic and mature by accepting her despair as stemming from her past and allowing herself to cry deeply, thereby releasing her pain. Suppressed anger is often vented on the person who tries to be helpful.

In Chapter 3 I emphasized the importance of getting a patient to cry, and showed that it is not as easy to do as one might think. Most children are not encouraged to express their sadness and some are actually beaten when they cry. As part of their training they have developed a stiff upper lip, and many pride themselves on their ability not to break down and cry even though they are hurting. Expressing one's sadness through tears and crying is one way of sharing feelings. Regardless of what some individuals say, most people respond positively to a person who cries. They may try to bolster his spirits, but rarely is he rejected because of his tears. But it is another story when it comes to despair and wanting to give up. We are like an army of stragglers trying to get home from a defeat and our chances of survival are threatened

by any weakness of will. "Keep trying, don't give up, plod on."
This would make sense if we were being pursued by an enemy
or, if nearby, there was a safe house awaiting us. But in this
world no one can find any real security other than in one's self.
Wealth, position and power are no answers to an underlying
feeling of despair and insecurity. In fact, it is the effort to overcome one's despair and insecurity that insures their continued
presence in the personality.

When Sally was feeling her despair, I suggested that she lie
over the stool and breathe. Then I asked her to scream "It's no
use! I'll never have anyone to protect and love me!" When she
did this, she broke into very deep sobs and suddenly found herself gasping for air. Her crying had stopped and all she could
say was, "I can't breathe, I can't breathe," and "I'm going to die."
But she was breathing and actually it was deeper than she had
ever breathed before in her therapy. True, she was gasping for
air but this represented a desire to live, not just to survive. The
gasping could also be understood as the result of the conflict in
her personality—to surrender to her sorrow and the fear that
she would be abandoned if she did so—or to struggle on. I supported her surrender, telling her to give in to the crying. As she
let go, the crying became softer and deeper. When she began to
gasp for air, her feeling was one of panic, but the panic disappeared as she gave in fully to the crying. I could see that her
chest had softened and her belly had relaxed. Then I suggested
she do some kicking and the grounding exercise to maintain her
deeper breathing. When she got up from the latter exercise, her
face had a different expression. There was a light in it I had not
seen before. Her eyes were bright. She said, simply, "I feel good."

The feeling of panic always arises in individuals when a strong
expiratory wave cannot pass freely through the diaphragm and
into the belly. It is blocked by a contraction in the diaphragmatic
muscle which can be painful and produce a feeling of nausea. It

is important to understand this reaction if one is to help patients breathe deeply. Nausea and the feeling that one will vomit develop when the wave comes up against the tension in the diaphragm, which acts like a stone wall causing the wave to rebound and move in the opposite direction—that is, upward. When the wave passes through the diaphragm and into the belly it enters the psychological underworld, a world of darkness. In mythology, the diaphragm, which is shaped like a dome, is conceived as representing the surface of the earth. But all life begins in the darkness of the earth or of the womb before it emerges into the light of day. We are afraid of the darkness because we associate it with death, with the darkness of the tomb and of the underworld. It is also the darkness of the night when consciousness dies and we go to sleep to be reborn fresh the next morning. The surrender of ego consciousness is frightening to many individuals, who have difficulty falling asleep or falling in love. Those individuals who are not frightened to death in their unconscious can descend into the psychological underworld of the belly and find the joy and ecstasy that sexuality offers. One must have the courage to confront the angel with the flaming sword that guards the entrance to the Garden of Eden, our earthly paradise, if one wishes to find joy.

Two weeks later when Sally came for her session she reported that she had lost the good feeling. I assured her that she would get it back if she expressed her sorrow and despair again. Lying on the bed and kicking strongly she screamed, "I'm tired of trying! It's no use! I can't do it any more!" Again the screaming opened up some deep crying, but this time she felt no panic as she let go into the crying. At the end of the session she felt the good feelings in her body again. Actually, the statement "I'm tired of trying" was relevant to her present life situation. She had often been required to work overtime and to take work home which interfered with her desire to spend more time with her son. Her neurotic personality did not allow her to protest. Submission was survival, which was all she knew. But as she became more alive through the deeper crying and breathing, she felt the pain of her situation more acutely, and also her anger about it. Strengthened by that feeling of anger she confronted her boss one day who, to her surprise, had no objection when she refused to work overtime except in an emergency.

Sally still had more work to do on her body. Its tightness had diminished noticeably, but it was still far from being full—that is, fulfilled. She could see the light at the end of the tunnel but hadn't reached it yet. She needed to continue the work, breathing to expand her chest more, screaming to open her throat more, and crying to soften her belly. This work would be ongoing for a long time to increase her sense of security and deepen her joy. She still had considerable anger to let out against her grandmother for frightening her, against her mother for abandoning her, and against her father for his seduction and rejection. Her relation to men was the critical element in her neurosis. Believing she needed them, she had allowed them to use her. At one point her anger against her husband exploded in a feeling that she could cut his balls off. But she recognized that her feeling of need had led to her acting seductively with men. That sense of need, however, had diminished greatly with the breakthroughs of strong feeling, which reduced the underlying panic and allowed her to feel that she could be alone and find joy in her freedom.

William is the golden boy whose problem I described in Chapter 5. I had worked with William for several years and he had made considerable progress in his life. He had been married many years earlier to an aggressive woman upon whom he was dependent. Then, when that marriage broke up, he became depressed. Mobilizing his energy, he lifted himself out of his depression and became active again in the world. He saw other women and began to move ahead in his profession, but felt frustrated that something was lacking. When he first consulted me, I could see from the enormous tension in his body that he was a tortured man. He could feel the tension and knew that he had to find some release, but while he accepted my statement about the severity of his tension he did not respond emotionally. He didn't cry and he didn't become angry. He was willing, however, to work with his body to deepen his breathing and become more grounded. This work helped him feel better and become more productive. At the same time, he worked on his relationship with his mother, who had made him believe that he was a superior being. The analysis took place simultaneously with the physical work. His father had never been a strong, supporting figure because his mother had taken exclusive possession of him. I now fulfilled the role that his father had abdicated, and he shared the events of his life with me.

Over the next several years William continued to make progress. He gained recognition in his profession and had met a woman for whom he felt love and respect. He also gained an ability to cry, which he did regularly in the sessions and at home when he did the bioenergetic exercises. He was by now quite successful and thinking of getting married to this woman. Just at this time, when everything seemed to be going well, he began to complain again of a sense of frustration. Despite his feeling of love for his partner, much of his sexual excitement with her had vanished. I mentioned in the earlier discussion of his case my belief that his inability to feel any strong anger against his mother was a major factor blocking his surrender. But as he still couldn't feel any anger, his frustration only deepened.

One session William came in complaining of a lack of enthusiasm for life, of a lack of passion for his wife or his work. Lying over the stool, he began to cry. I suggested that he say, "Oh, God, it's such a struggle." His throat tightened and he couldn't get the words out. He got up, saying "There is something frightening about that." He looked frightened, almost in a state of panic. I asked him to lie back over the stool again and say, "Oh, God, I can't get enough air." He did and then added the words, "It's true." He felt a fear that was between panic and terror which he had never allowed himself to experience before. Then he told me a most important detail of his life: Every month or so, as a boy, he would cry for several nights when going to sleep. "When I would wake up it was with a very bleak and dark outlook for my future," he said, "but then when I got out of bed and active, it left me." He also admitted that he still felt this bleakness now, but that it was short-lived.

William had been in the grounding position as he related this. When he got up, I was amazed at the change in his face. It was soft, bright and young-looking. It was as if he had been released from a dark cell. I realized then that his habitual expression was a mask. He often smiled but the smile was as hard and tight as his body. The change in his face now was due to the acknowledgement of his despair. "I am disillusioned with my life," he remarked. But why did he have to hide it, even deny it? This denial betrayed the existence of a profound fear.

As we talked about his lack of anger, he said, "I see myself as a social success. I have money, friends and possessions. I don't feel any worse than other people." It was clear to me that he was deeply ashamed to show that he was in bad shape. He was raised to believe that he was a superior person, godlike. He could not be one of the common people. He was forbidden to show any sexual interest in girls. "Sex wasn't acknowledged in my family," he said. "My mother never said a word about sex to my sisters. She spent a lot of time in church. I was an altar boy. She was preoccupied with cleanliness and godliness, with being clean and being good." When William didn't obey, he was reprimanded, and when he was "bad," he was spanked. He was never beaten.

What, then, was the great fear that had forced him to deny his feelings and strive for superiority at all costs? Then I realized that William's mother had a streak of insanity in her personality, as all fanatics do. As a boy he was terrified of what she could do, and was in a state of panic that she would reject him if he challenged her. I had alluded to her fanaticism as a sign of unreality in her personality throughout the therapy, but William saw her only as unusual. Now, for the first time, he could accept that there was a streak of insanity in his mother. The blinders were lifted from his eyes and he could see some light. The world was no longer a bleak, dark place. That light grew brighter in succeeding sessions.

Mary's story reflects the steps in her therapy which led to the considerable progress she made towards joy. Briefly discussed in an earlier chapter, she was a Gestalt therapist, thirty-three years of age and married when she started her therapy with me. She had attended a professional workshop I did for a group of therapists and was greatly impressed with my ability to understand her struggle by analyzing her body. Its outstanding feature was a marked split between the upper and lower halves of the body, which looked as if the two halves had been pulled apart. Her waist was thin and elongated. Both halves looked weak; her chest was tight and contracted, her neck was thin and slightly elongated and her face was soft and weak-looking. The lower part had a similar look of weakness, manifested in a narrow, tight pelvis and thin, long legs. Her feet did not appear strong either. The look of weakness in Mary's body denoted a reduced energetic charge which was also manifested in a diminished intensity of feeling. Her self-assertion, for example, was weak. In addition, her body showed a lack of integration between its parts; head, thorax and pelvis were not well connected energetically to each other.

When I pointed this out to Mary at the workshop, and remarked that she had a big problem which required a body-oriented, therapeutic approach, she told me that no other therapist had seen her difficulties. She had a doctor's degree in psychology and, on a verbal level, she could handle herself very well, which fooled most therapists. She had an attractive, young face and an eager smile which expressed her desire to please but also covered her sadness and panic. When we started to work, she was thankful that I saw her pain and sadness. She welcomed the encouragement to cry which she desperately needed to do. She also did some kicking and screaming, using the word, "why" to protest what she knew was an unhappy childhood. It was not difficult for Mary to sense how badly hurt she had been as a child. As we worked to increase her sense of herself as a person, she related memories and incidents from her childhood which showed how frightened she was. "When I was little my mother used to tie me up. She once tied me to the outside screen door. I remember screaming and screaming to be let in but she ignored my cries. Both my sister and I used to be hit and beaten by my mother with a wooden spoon or a coat hanger."

Mary recalls her childhood as a nightmare. She used to walk in her sleep as a child, and sometimes would run as if trying to get away. She had frightening dreams. "I would be in the sea and sharks were coming at me. Sometimes I would wake up before they attacked me but other times they would bite my leg off before I could wake up. There was blood in the water. I don't remember screaming but I would wake up in terror. There was another type of dream that was less clear. I was in a woods and a serpent was after me, but I felt paralyzed and couldn't move away. Those dreams occurred between the ages of four and five. Even now I can sense a terror in me. I was a very anxious child but I pretended to be brave. Even at twelve years of age I felt terrified if I had to ask for something from someone. It was torture for me."

When I asked Mary who she thought was the shark, she said, "I always thought it was my father. Lately, however, I sense a lot of fear in relation to my mother. I never had a sense that my mother hated me. Now I feel that my mother doesn't love me. I am afraid to face the fact that she hates me."

During this session Mary revealed that she had known that her parents had married because her mother had become pregnant with her. It was her feeling that her father hadn't wanted her mother. There had been a conflict over her name when she was born and she was finally given the name that was her father's preference. Then she said, "When I was young I always felt I was my father's real bride." Mary was aware that her father was sexually involved with her although she had no sense that he abused her.

It was essential for Mary to feel her body problem and how it was caused by her childhood experiences. In the therapy I kept her focused on the split in her body and her need to integrate its segments. This is accomplished by getting the wave of excitation associated with breathing to flow strongly through the whole body. Breathing over the stool promotes this flow. In one session, lying over the stool and breathing, she began to cry and said, "Oh, God. I can't stand the split in my body between the upper and lower halves. I feel like I'm on a torture rack." She had been tortured psychologically and her body had been broken by the emotional conflicts in her home created by her father's sexual interest in her and her mother's jealousy and hostility toward her. At the same time she couldn't protest against what was happening to her because her parents were blind to their behavior. With my encouragement, she screamed, "You're torturing me, and I can't stand it!" But then she added, "I feel like I can't get away!" With that remark, she collapsed on the floor, sobbing deeply.

She added, "My mother was after me all the time, attacking

me every time I attempted to be free or to show any sexual feeling. I gave up. I became her little servant girl and she was very happy. But I was very awkward afterward in school. I thought something was wrong with me. I felt guilty about my anger toward her." But she was also guilty about her sexual feelings for her father. In a later session she complained about a feeling of agony in her pelvis. She sensed a reluctance to go more deeply into that feeling. Then, as we discussed her fear of feeling deeply into her pelvis, she said, "Oh, God. I sense that I am holding against the insanity of my father. He would go crazy if I allowed my sexual feelings to come through." She began to cry deeply, then added, "It feels like all the energy of my father is coming at my pelvis. His eyes were always looking at my pelvis. It was crazy—tormenting—unbearable. I knew he was perverted but now I feel it directly. But since no one confirmed it, I was made to feel the bad one. I cut off the sexual feelings in my pelvis and I became an 'angel,' a good Catholic girl. When I felt any sexual feeling or showed any sexual excitement, I felt perverse. It's pretty sad. But," she added, "I have good body feelings now and while I feel more sexual, I am less seductive in my behavior." This improvement was the result of the release of tension in her body through the crying, the screaming, the kicking and the hitting, which allowed the wave of excitation to flow more freely. She also did regular bioenergetic exercises at home, which strengthened her body. As a result of the strong body work and the concomitant expression of feeling, her fear diminished greatly.

At one session, as she lay over the stool breathing, I stepped out of the room for a minute. When I returned she was in a state of panic. She cried out, "Don't leave me with her." When I asked what she was afraid of she said, "I feel she would rip out my vagina." It was clear to both Mary and me what her struggle was. Feeling hated by her mother, she turned to her father for love but his love had a perverted quality which excited and frightened her and, at the same time, rendered her more vulnerable to the jealousy and rage of her mother. She had been literally pulled apart by her two parents, each of whom demanded a different emotional pattern. Her mother demanded an asexual, virginal quality while her father responded to her sexuality.

In one session, when she was over the stool, she sensed her difficulty in breathing. She had been crying and her throat had contracted. She said, "If I cry too hard, I'll choke to death. I'll die." But she couldn't stop crying. "Oh, God" she said, "My sadness is overwhelming. I can't stand it. She hates me and I need her. My chest feels like one great scream against her cold, hateful eyes. Oh, my God! Without my father's love, there would have been no reason to live. That's why men have been so important to me."

Mary had pulled away from her sexuality to avoid being overwhelmed by her father's sexual interest in her and to protect herself from her mother's jealousy and anger. But this action destroyed her integrity and undermined her security. In her vulnerability she turned to men for protection and love. The result was that men used her sexually in the name of love, which further undermined her sense of self. To become more independent, more self-assertive, she needed to see her self-betrayal. She remarked, "I am shocked that I can be so sweet and giving to men. I have always felt special to my father, to my professors. If a man makes me feel special, I give him sex." At the same time, she was able to mobilize her anger against them for using her. However, because of the destructive actions of her parents, there was a murderous rage in her which had to be released slowly. It was too frightening.

Mary's relation to men was as twisted as her relations with her parents. On one hand she felt special, on the other she was angry. She said, "They act as if they possess me, which makes me furious. But I also feel guilty toward them, which I recognize is a denial of the feeling that I want to hurt them." Her self-awareness deepened with each session. "I realize that I let myself be a victim, letting other people take out their hostility and bad feelings on me. Before that realization, I saw myself as an angel." Later she added, "I want people to make it up to me, to take care of me. I feel that I've been such an angel and they owe me something." She also realized how neurotic this attitude was, and she could sense her rage and feel its murderous quality. But hitting the bed with the tennis racquet and saying, "I could kill you," frightened her. She remarked, "I could feel the madness in me." Then, as she accepted the feeling of anger/madness, the fear diminished. And as her anger became stronger, she said, "With this feeling I don't need a man to protect me."

Hitting the bed during another session, she remarked, "I could feel the warmth move up my back as I hit. It feels good to have a back (backbone feeling) and a front." The emotion associated with the back is anger while the feeling along the front of the body is longing and love. Mary could now understand how and why she had lost the feeling of having a backbone, of being able to stand up to people. "When I got angry as a child, my father became enraged and my mother blamed me. Reading my diary I saw how I suppressed my anger. If I got irritated by someone, I would blame myself. I wanted to be good. That was my mother's idea of how one should be. My father was an angry man and I didn't want to be like him. When I was young—seven to nine—I would feel guilty if I was fresh with my mother and I would confess to the priest."

There was another aspect to Mary's therapy which promoted her self-esteem and self-possession and that was the focus upon feeling connected to her pelvis and her sexuality. This was done by increasing the charge in her pelvis through deeper breathing and deeper crying, which caused the lower part of her body to vibrate strongly as the excitation flowed downward. Doing the grounding exercise described above also helped greatly. The release of any strong emotion increases the flow of excitation. In one session, after she had kicked strongly on the bed while screaming, "I can't stand it! I won't stand it!" her pelvis began to move with her breathing. She remarked that she had some very nice, pleasurable feelings in the lower part of her body. This feeling persisted for two weeks, during which period she also felt tired, which stemmed in part from her moving to a new home, but largely from the giving in to the body.

Struggle is tiring and the struggle for survival is very tiring. Since most people in our culture are survivors, fatigue is the most common symptom in the population. It is the physical side of feeling depressed. But survivors cannot afford to feel tired or depressed because they would be tempted to give up the struggle and die. Their defense is to deny the fatigue and carry on because they sense that their survival depends on it. As one woman said, "If I lay down, I feel I'll never get up." But until one is ready to lay down, one denies the feeling of tiredness. A traveler carrying a heavy suitcase and running for a train will not feel how tired his arm is until after he lays down his suitcase. In therapy, to feel tired is a mark of progress if one can associate it to giving up the struggle.

When Mary came to the next session, she remarked that she felt more womanly. I observed that she was more in contact with herself and her body. She described her feeling as an inner quietude which she had not experienced for a long time. I noticed that her voice was deeper and that there was a complete absence of anxiety in her behavior. Lying on the bed, she said "There is a warmth moving up my lower back from my pelvis. It is very pleasant. I feel a soft sadness and I want to cry. I feel that I am returning to myself. I feel at home." She mentioned that as her pelvis was moving spontaneously, she felt her lips also beginning to move. "They feel connected to each other," she said. She was now crying softly and more deeply.

"I was thinking about my father and the men I've known. I could sense the pain of their loss but at the same time I have a good feeling of myself, of being separate. As I separate I have such a wonderful feeling of myself. It makes the separation worthwhile. I sense that when the feeling of being separate becomes too strong, my pelvis pulls back and the old feeling, 'Daddy, I need you,' arises."

"I feel I have to make a choice between men and me. I can't be there for them and also be there for myself." As we discussed this issue, I pointed out to Mary that when she focused on her sense of self rather than on what a man could do for her she was truly a sexual woman. When she used sex to get love from a man, she took the role of the daughter/prostitute. A sexual woman can contain her sexual excitement instead of needing to discharge it.

Mary remarked, "I feel like a different person, like being reborn." She began to cry, saying, "I've always longed for this."

This breakthrough didn't mean that Mary's therapy was finished. In her voyage of self-discovery she had traversed her inner hell but purgatory was still ahead for her. Considerable work needed to be done to strengthen her connection with her sexuality and her pelvis. This connection was associated with despair. "If I am sexual, I can't have my father. If I have myself, I can't have a man." Mary was smart enough to realize that this either/or didn't make sense, that being for one's self doesn't mean that one can't have a partner, but knowing this intellectually didn't change her feelings. The split between ego and sexuality was deeply structured in her personality and in her body, and associated with it was a deep feeling of despair, against which she was still struggling. But she was close to surrender to her body.

The surrender to the body is basically a surrender to its sexuality which lies at the very base of the body structure, namely in the pelvis. This surrender is a physical process, not a psychological one, although it can be aided by an understanding of the fears which block the surrender. The physical process of surrender involves allowing the wave of excitation associated with breathing to flow into the pelvis and into the legs. When this happens the pelvis moves spontaneously, backward with inspiration and forward with expiration. This spontaneous movement was called by Reich "the orgasm reflex" because it occurs at the climax of the sexual act when an individual surrenders fully to his sexual feeling. The result is a feeling of joy as pointed out in the preceding chapter, where I described the exercise known as the pelvic arch.

This exercise can also be used to promote a feeling of self-possession in the individual. Self-possession, in my view, entails the right to have and to hold: to have one's self and also to have and hold a loved one. When the patients' pelvis becomes charged as they do this exercise, I place a rolled-up blanket between their thighs and ask them to squeeze it as tight as they can. I also suggest that they thrust the lower jaw forward to mobilize aggression. Sometimes I ask the patient to bite on a rolled up hand towel at the same time. Focusing the aggression upon an object strongly increases the charge and the vibratory activity of the pelvis which spreads into the legs and feet. Of course the towel can be seen as representing the breast while the blanket represents the body of the loved one. If one can allow the feeling of possession to take over, one gains a strong sense of self-possession as well as the feeling that one has the right to possess the world. This allows an individual to unite with the world or universe in an active rather than mystical sense. It becomes a basis for a sustained feeling of joy.

As she came in for another session, Mary reported "I really

feel happy now. I have these sweet feelings for some men but I am not hung up on them. I enjoy the feelings. I can be alone and feel good in myself. I have both the feelings and the freedom, which is wonderful."[3]

Then she added, "I appreciate your help and your not being involved with me. It allows me to be free and not involved with you."

As long as people are involved with each other, they are not free. Needing something from each other, they are dependent. Dependency in a relationship throws each individual back into his childhood experience, where he was dependent and vulnerable. To free them both from dependency, to help them grow into mature adults, one needs to understand the role of sexual guilt in making a person become submissive—that is, being there for others. The concept that each should be there for the other is a business arrangement in which neither individual is there for himself. How sexual guilt operates to create a neurotic character is analyzed in the next chapter.

[3] See Lowen, A., and R. L. Lowen, *The Way to Vibrant Health*.

CHAPTER 11

PASSION,
SEX AND
JOY

In the previous chapter I discussed the fear of death, which I believe lies at the base of all the emotional problems which people bring to therapy. The fear of death results in a fear of life. One cannot surrender to life or to the body because surrender means letting go of ego-controls, which would bring one face to face with the fear that he would or could die. That fear stems from a very early life experience of coming close to death or the possibility of death, which causes the organism to armor itself as a defensive measure so as not to be vulnerable to that possibility again. But living in a state of being armed or armored means that one accepts the possibility of being attacked or threatened with the loss of life. This is the psychological and physical condition of the survivor. The energy that is invested in the effort to survive is not available for the enjoyment of life. But this also means that a person's fear of death prevents him from living fully and brings him closer to death.

Life and death are opposite states. If one is alive, one can't be dead and vice versa, but one can be half alive and half dead as we noted in the preceding chapter. If a person is not fully alive, he is partially dead and consequently is frightened of death. The fully alive person is not frightened of death because he is not frightened. He is free of the chronic contractions that represent fear. His body is loose and relaxed. Such a person doesn't deny death, but it is not a physical reality until it occurs. When it does occur, he is not afraid because there is no feeling in death. Life is the antidote to the fear of death.

A brave man is not afraid of death for that is the essence of bravery. Throughout history men have risked their lives for liberty and freedom because liberty and freedom are essential if one is to feel joy. Without freedom joy is impossible and without joy life is empty. During the debate in the Virginia assembly on the question of independence from England, Patrick Henry uttered words which have now become famous: "Give me liberty or give me death." His feeling for liberty amounted to a passion which was strong enough to counter a fear of death. Other brave men have acted similarly because they too had a passion which was strong enough to enable them to face death without fear. Many people have died for their religious beliefs because those beliefs were connected with a passion for the principles or doctrines of the religion. But lovers have also risked and lost their lives in the pursuit of their passion. It is the nature of passion that it will move the individual to actions that transcend the ego's drive for self-preservation. Only in this transcendence can an individual experience the joy and even the ecstasy that life offers.

True passion, by its very nature, is life-positive even when it can end in the death of the individual. It seeks the enhancement of life. We speak of a passion for art, for music, for beauty when these aspects of life arouse strong feelings in a person. We would never speak of a passion for alcohol, for gambling or for any act that is destructive to life. One can get passionately angry about an injustice but going into a rage is not a passionate feeling. The difference, I believe, lies in the fact that passion is hot; it stems from an intense fire. Anger is also hot. But rage is cold, even though it is violent. Many people have strong feelings of hate but these feelings do not constitute passion. The hot feelings are related to love, and that includes anger, as I showed in Chapter 5.

We all know that sexual feelings can reach a level of passion if enough love is associated with the sexual desire. Sexual desire results from an excitation of the genital apparatus, whereas the feeling of passion is located in the pit of the belly as a warm melting sensation. The genital excitation can reach a high intensity, but when it is limited to the genital organs it does not qualify as passion, in my view. The need to urinate or evacuate can also become very strong and result in feelings of pleasure and satisfaction when it is fulfilled, but those limited sensations do not constitute passion. Passion, along with love, anger and even sadness, are emotions, which means that they embrace the whole body in the feeling. Sexual desire is an expression of love since it aims to bring two individuals together in the mutual experience of pleasure. But when the desire is limited to the sexual contact, it is too narrow and limited an expression of love to constitute passion. Under such circumstances the sexual act does not result in the feelings of joy and ecstasy which it can provide.

The split between sex and love, between sexual desire and sexual passion, is connected to the split in the personality between the ego and the body. If the ego does not surrender to the body in the feeling of sexual desire, the act of sex becomes a limited expression of love and, therefore, unfulfilling on a deep level. This inability to be fulfilled in love on a sexual level maintains the feeling of despair which the individual experienced in his early relationships. I believe we must be critical of the modern sophisticated view that the act of sex in itself provides a sense of fulfillment, or that the ability to function sexually is a valid criterion of health. In this culture we are preoccupied with performance with no regard for the feeling essential to making any act an expression of health.

Adults cannot experience joy as children do because the adult position imposes responsibilities for actions and behavior from which a child is free. Thus adults cannot play in the carefree way of children. Adult play always has a serious note because the adult ego is involved in the outcome of the activity. For example, playing a game of cards which is fun for children is a serious undertaking for adults for whom the outcome, winning or losing, is often more important than the game itself. When winning or losing becomes important in children's play it is a sign that their egos have developed to the point where they are self-conscious. The self-conscious ego judges and controls behavior which destroys the individual's ability to surrender freely and fully to his feelings. This does not mean that adults cannot experience joy. They do not experience it in their ordinary activities, which are serious, since such activities are largely concerned with earning a living or protecting life. A healthy adult can experience such activities as pleasurable if he can willingly accept the responsibility for their outcome. There is, however, one activity in which the surrender of the self-conscious ego is encouraged and that is the act of sexual love. The loving surrender in the sexual act results in an orgastic release which embraces the whole body in its convulsive movements and which is experienced as joyful and even ecstatic. Because this orgastic response is deeply fulfilling the feeling of joy persists, giving life its deepest meaning.[1]

The ability to experience a total orgasm is the mark of a passionate nature. It is the result of the buildup of a level of positive

---

[1] Reich, Wilhelm, *The Function of the Orgasm* (New York: Orgone Institute Press, 1941); Lowen, Alexander, *Love and Orgasm* (New York: Macmillan Publishing Company, 1965).

excitation strong enough to overwhelm the ego and allow the person to express the full passion of his (or her) love freely and totally. In such an orgasm there is no ambivalence, no holding back, no hesitation in the surrender of the self. Some individuals have experienced such an orgasm on rare occasions. Unfortunately, it is not a common experience. For very few it is their normal sexual response. There are individuals who have a truly passionate nature who are able to commit themselves mind, body, and soul to their feelings and actions. I have known such individuals and they are a joy to be with. This is not to say that their every act is intense and passionate. Rather, they have a capacity for passion which is manifested in the brightness of their eyes, in the aliveness of their bodies and in the gracefulness of their movements.

Unfortunately, in childhood, it is this aliveness which creates the conflicts that lead to the suppression of passion. The preceding chapters of this book detail some of the traumas and fears which undermine the integrity of the child and forced him to suppress his passions. While those conflicts generally start early in life they reach their climax during the oedipal period when the love of a child for the parent of the opposite sex reaches its first sexual flowering. As in the oedipal myth the feelings that are aroused threaten the child and sometimes the parents. A child in this situation feels that his life is threatened unless he withdraws from the situation by cutting off the passionate sexual feelings he had for the parent of the opposite sex. The cutting off of the strong sexual feelings amounts to a castration. I have found this fear of castration in all patients, where it is associated with a fear of being killed. This resolution of the oedipal situation results in a diminution in the strength and intensity of all feelings, in sexual guilt which is unconscious and in the development of a characterological, or habitual, attitude of submission to authority. This attitude may be countered by a superficial

rebelliousness, which is an effort to deny and overcome the submission.

Restoring a person's ability to feel passionately is the therapeutic task as I have described it throughout this book. It involves
energizing the body through deeper breathing, encouraging the
patient to cry more deeply, helping him understand the origin
of his fear and removing it through the expression of anger in
the therapeutic situation. The aim is to help the patient feel that
he is free to express himself in an appropriate manner. But the
key to passion is the recovery of full sexual excitation especially
in the pelvis not just the genitals. This can only happen when
the flow of excitation connected with breathing extends into the
pelvis, integrating the segments of the body so that head, body
and soul are experienced as a unity.

For a patient to find his sexual passion he needs to get more
energy and excitation into his pelvis. He also needs to understand
the fears that block this downward flow. Mary was a patient
whose case I discussed in Chapter 10. She had gained a good
understanding through analysis and body work, but her fear of
her sexuality was still considerable. In one session she described
a breakthrough as follows: "When you pressed with your fingers
upon the muscles in my pelvis and I breathed way down into
the area of your pressure, I had a feeling of paradise. But I
couldn't hold it and I felt sad and cried." In this procedure the
person brings their energy down into the pelvis to alleviate the
feeling of pressure. The result is a sensation of aliveness and
fullness in the pelvis. But she could not sustain the good feeling
on her own because she was still too frightened and ashamed.

Mary came into a succeeding session with a different attitude.
She remarked, "I'm fed up with being so anxious, so afraid. I
don't want to go on this way anymore. I'm tired of the struggle.
I'll just take life as it comes. I'm sure I'll survive." She was closer
to a surrender to her body. This new attitude stemmed from a

deeper, more painful insight. "I've never felt how crippled I am, how much I've been devastated. I feel the shame so strongly, I want to cover my face." It was a shame of sexuality. She added, referring to her father, "I was always his little woman-child. I felt so special, great. Then it burst and I felt I was nothing, just a piece of dirt."

Mary said, "When I get through the shame, I feel my eyes become bright. It's such a lovely feeling. I feel a softness that is sweet, a melting sensation. Oh, God! It feels so sweet in my pelvis but my head is crazy." More work was needed so that the flow of excitation upward and downward would remain anchored in a clear head and a softly charged pelvis. That can only happen when the fear of surrender is fully resolved.

Mary came to her next session after having been to a weekend workshop. She began by saying that she felt a resistance to coming and that she was reluctant to open herself to any feeling. She related that she had worked with a woman therapist on the issue of her relationship with her mother and had cried as she experienced a longing for her mother. Then she described how on the way home from the workshop she sang what she called a little mommy song, as if she were a little girl. It was obvious to me that she had regressed, abandoning the more mature position she had achieved. Such a step backward indicates that she had touched upon a deep fear. This was borne out by a dream she had immediately following the workshop. She related that she was with a girl in the dream who was trying to kill her with a knife. She felt that when the girl tried to stab her in the heart, she could protect herself but then the girl made a move to attack her pelvis and she felt helpless. It was as if the girl would kill her one way or another. When I asked Mary whom she thought the girl was, she answered immediately, "my mother." She then related that she always felt that her mother had not wanted her because she was a girl. Sensing her mother's hostility, she had turned to her father for acceptance and love, which he gave her. But it became perverted by his sexual interest in her. In her innocence as a little girl, she accepted his interest and affection wholeheartedly, which saved her, but at the same time she was betrayed by it. She did not realize the betrayal until her illusion of being special and pretty collapsed in the face of the humiliation she suffered when he exposed her to his drinking friends. In desperation she gave up her sexuality and turned to her mother and to the church, becoming a devoted daughter and a very religious Catholic. But she still felt ugly and full of shame.

This devastation could not have occurred if her mother had been there for her. If she had had her mother's love, she would not have lost herself to her father, becoming his woman-child. The relationship of her parents to each other was a distorted one. Her mother was cold, tight, very religious and anti-sexual. Her father was sexually loose, good-looking and pleasure-oriented. Opposites attract. These two people were drawn to each other because each needed what the other had. But since they could not accept and surrender to that need, they attacked what the other represented. Mary became the victim, the one in between who took the beating—especially from the mother who envied her and hated her for the sexual excitement she had with her father. Mary felt so guilty about her sexual involvement with her father that she was lost and helpless. Her fear of her mother had destroyed her integrity as a person and that fear was still present in her. To feel solid in her growth and sexuality, she had to face the fear and find release from it through the mobilization of her anger. She understood my explanation of the situation. Lying on the bed and twisting a towel with her hands, she opened her eyes to look at her mother and said, "You really hated me, didn't you?" Saying this, she saw her mother's face and the look in her eyes which frightened her. She said, "I get frightened when I look into someone's eyes, especially those of a woman. For years

I couldn't see my mother's eyes. Then, when I was grown, I recalled a picture of her that I saw as a child of four. I remembered those cold eyes that looked as if she wanted to kill me. I felt paralyzed. I couldn't breathe."

To help Mary resolve the fear I had her reverse the exercise. Twisting the towel, she screamed at her mother, "I hate you. I could kill you." Expressing these feelings, she remarked, "They make me feel pretty. I used to feel so ugly." And with anger, she added, "Don't look at me that way. It scares me so much." Mary had never before mobilized a murderous anger against her mother. She had felt too guilty about her sexual involvement with her father and too frightened of her mother. It had taken almost three years of therapy to bring her to the point where she was sufficiently released from her feelings of guilt and shame to be able to stand up for herself. She had gained strength and confidence in her ability to survive alone, to stand on her own feet. But it would be wrong to think that this breakthrough marked the end of her therapy. Terms like stronger and more self-confident are relative. Her body needed considerably more work to increase its energy and to further the integration. In the face of sufficient stress or disappointment in her relationships, she might still collapse. We never fully overcome the effects of life's early traumas. But should we be hurt again, we can more quickly mobilize our forces and restore the state of good feeling and pleasure in our body. Each crisis that we meet in life becomes an opportunity for further growth in our selfhood. In effect, then, the therapeutic process is unending. Our voyage of self-discovery is never finished as long as we live, since each life experience can add to the richness of our being. That has been true for my personal voyage.

I had been drawn to Reich by his thesis that one could find sexual fulfillment through the surrender to one's sexual feelings. He named this ability orgastic potency, to denote that sexual passion was not measured by how strong one's sexual drive was but by how full and complete was the discharge or release of the excitation. In a full or complete orgasm the whole body, including the mind, participates in a convulsive reaction which completely discharges all sexual excitation. That convulsive reaction is triggered by waves of excitation that pass through the body connected with the heightened breathing tempo. Though I use the term convulsive as Reich did, the movements are not chaotic nor clonic; they are snakelike. In this action the pelvis moves forward with expiration and backward with the respiratory wave. The same movement can occur with deep and full breathing without any sexual charge or genital excitement. In this situation, the movement is called the orgasm reflex and does not build to any climax. It is experienced as very relaxing and pleasurable. In the sexual act, when the strong sexual charge in the genital apparatus explodes, the pelvic movements become completely involuntary and are fast and forceful. One feels carried out beyond the self which is the highest form of surrender. The consciousness of self disappears as one feels a merger with cosmic processes. The experience is one of ecstasy.

As a result of my therapy with Reich I was able to experience the full surrender to my sexual feelings and know its ecstasy. It has, however, been a rare experience. Nevertheless, it has strengthened my conviction that love and sexual passion are aspects of man's identity with the universal. But if that identity is part of man's nature, why is it so difficult to surrender? I have described the fears that prevent or block that surrender, but since they are universal fears in our culture, we must recognize that they have a direct relationship to that culture. What happens in the family reflects cultural attitudes and values, and unless we recognize the distorted nature of these values, we are helpless to avoid this destructive effect on ourselves and our children.

Culture developed as man moved out from the purely animal

state and became a self-conscious individual. That move upward from the four-legged stance of all other mammals to upright posture lifted man above the other animals and also, in his mind, above nature. He could observe the processes of nature objectively and learn some of the laws which governed their action. And in doing so he began to gain some control over nature, and by extension over his own nature. He developed an ego, a self-aware and self-directing agency which enabled him to gain an ascendancy over the other creatures, which led him to believe that he was different, which he certainly was, and special, which he wasn't. This development was made possible by an evolutionary step through which man gained a more highly charged body and a greater range of physical movement, especially in his hands and face, including his vocal apparatus. He can do more things and express himself in more ways than any other animal. In these respects he is superior to them but still not special. He is born like other animals and dies like them. His feelings may be more subtle but they, too, feel. He has flourished and achieved much in his short sojourn on this earth, but his upward progress has alienated him from his base in the earth and in nature and his activities have become destructive to himself and to nature. The destructive impact of our culture on nature is now fairly well accepted, but we are not ready to recognize the destructive effect it has upon the human personality. We see the damage in child abuse, rampant violence, depression, addiction and sexual acting-out, but we believe that it is within our power to control and remedy the situation if we have the will to do so.

My thesis is that the will is impotent to change this state of affairs because the will is part of the problem. We have gained power and we are hung up on it. Our culture is driven by power, literally and psychologically. Without power our civilization would come to an end, but as power increases, it is driving us faster and faster in all our activities to the point where we are losing control of our lives. Our bodies cannot keep up with the pace of the activities demanded of them—which is the basis of stress. If we relax for a few minutes it is only so we can run faster for the next few minutes. We are driven to keep up, we are driven to succeed, we are actually being driven out of our bodies. In the more than fifty years since I began studying the human condition, I have seen a general deterioration in the bodies of the people who come to me. They are less energized, less integrated and less attractive than the bodies of the patients I used to see. Borderline conditions are almost the dominant disturbance. The old-fashioned hysterical patient that Freud wrote about is almost never seen. The hysterical person couldn't handle his feeling; the schizoid individual hasn't many. Most people today are dissociated from their bodies and live largely in their heads or egos. We live in an egotistic or narcissistic culture where the body is seen as an object and the mind as the superior and controlling power.

In the context of the therapeutic process, power and the will are the negative forces that impede healing. The power is in the mind of the therapist, for he sees himself as the agent who can produce the desired changes in the patient. He may know consciously that he cannot change the patient, but his knowledge of the psychology underlying the patient's emotional distress can give him a sense of power if he, like most individuals in this culture, is narcissistic and has a need for power to support his self-image. That power is exercised through his judgment and control of the analytic material. In one way or another he can indicate his approval or disapproval of what the patient says and does. And since he is the guide who must lead the patient through the underworld, he does have this power. So does every parent. If a therapist denies this power, he is out of touch with the realities of life. The issue is whether he recognizes and accepts that he has power and does not let it go to his head.

Power is the issue that I struggled with throughout my therapeutic practice. With the ability to see a patient's problem clearly by reading the language of his body, I believed that I could direct him as to what he must do to get better. When the patient did what I directed he generally felt better, but it did not hold up. Though I had learned from Reich that the issue is not doing but feeling, my own personality was such that I could not refrain from trying to *make* it happen. I must have believed that if I could make it happen, I would be the super individual I was supposed to be. I believe that almost all people in this culture have been indoctrinated with the idea that one has to try to make it happen—that is, to become healthy and potent, successful and loving. I know that it is true of my patients, and I have realized that it is equally true of me. If what we are looking for is passion, sexual fulfillment and joy, we cannot make it happen any more than we could make life happen through our will and our trying.

When I work with people now, I am still in control of the therapeutic process because I am the guide. It is my responsibility to understand my patient and his problems and to point them out to him so that he can see and understand them too. Without my understanding we are both lost; without his understanding himself, he is lost. It is my responsibility to guide him in his voyage of self-discovery. But the healing is beyond my control.

Healing is a natural function of the body. If we cut ourselves, does our body not heal spontaneously? Living organisms would not have survived this long without the innate ability to heal their injuries and illnesses. As doctors we can help the natural healing process but we cannot do it. If this is so, why don't we innately heal our emotional disturbances, since they represent injuries to the body as well as to the mind? The answer to that question is that we don't allow the healing to occur. We block it consciously and unconsciously out of fear, as we have seen in the preceding chapters. We cannot remove our fear by a deliberate act of will; all we can do is suppress it so we don't fear the fear. But as a consequence we suppress the body's vital activities, including the process of natural and spontaneous healing. It is only by the surrender of ego-control that one's body can regain its full vitality and energy, its natural health and its passion.

The surrender to the body and its feelings may strike one as a defeat, which it is for the ego that seeks to dominate. But only in defeat can we gain freedom from the rat race of modern life to sense the passion and the joy that freedom offers. But this goal is not easily achieved. We are burdened with the knowledge of right and wrong and with a self-consciousness that limits our spontaneity. And as I have pointed out elsewhere, the voyage of self-discovery is never finished. Therapy, however, is a practical issue. One can't and shouldn't be in therapy all his life. Six years should be the maximum, since it takes that long for a child to gain enough independence to leave home for school.

When a patient terminates bioenergetic therapy, he should have available the understanding and techniques that will allow him to further the process of self-awareness, self-expression and self-possession. He should understand the connection between the body and mind and know that his chronic tension is connected to unresolved emotional conflicts stemming from childhood. These conflicts operate in the present as long as the tensions persist in the body. He will, therefore, work with his body to reduce them, even to eliminate them. This means that he will continue to do the basic bioenergetic exercises as part of his normal health-care routine. I do them almost every morning as regularly as I brush my teeth, and I have done them for more than thirty years.

For breathing, I lie over the bioenergetic stool for three to five minutes, allowing my breathing to deepen. To further this process I will also use my voice, making and sustaining a loud sound. Though loud, it is an easy sound made without effort. Generally, the effect is to induce some sobbing. Once I start crying, my breathing becomes easier and deeper. Crying is important for me because I have always had a resistance to crying, for all the same reasons that others resist crying. I have been a determined person trying to rise above my problems. Though it hasn't worked, I've been unable and unwilling to give up. Crying is a giving-up and that meant failure. But giving up is what therapy is about and I have learned over the years that each time I give up in any area of my life, I gain freedom. But my neurotic character is so deeply ingrained in my personality that it is a continuous process. I give up only a little each time.

Crying serves another related function in my life: It keeps me in touch with my sadness—the sadness of years when I wasn't free to be true to myself and the sadness that I will never regain the state of innocence that would be pure joy, or what is called bliss. In contrast to the animal, we live with the knowledge of struggle, suffering and death. It is the tragic side of the human condition. But the other side is to be able to experience the glory of life in a way no animal can. In religious terms it is referred to as the glory of the Lord. I see the two as synonymous. That glory is seen in the beauty of a flower, a child or a woman, and in the majesty of a mountain, a tree or a man. The experience of that glory is an exaltation which finds expression in man's artistic creations, especially in music. It is a basic thesis of my philosophy that one can't separate the two sides without destroying the whole. One cannot experience the glory if one cannot accept the tragic aspect of life. There is no glory if one denies or escapes from reality. I need to cry to retain my humanity. I cry not only for myself but for my patients and for all mankind. When I see the struggle and the pain in my patients it often brings tears to my eyes. Then, when they release the pain by crying and give up the struggle, I see their eyes and faces light up and my heart rejoices for them. But I can only feel this joy if I, too, am prepared to give up the struggle, which is why I need to cry.

Another exercise that I have done ever since I created the bioenergetic approach is the grounding exercise. After I work out on the stool to deepen my breathing, I reverse the position by bending forward and touching my fingers to the floor. This exercise is fully described and illustrated in Chapter 2. Holding this position will generally cause my legs to vibrate as waves of excitation flow through them. The vibration not only deepens my breathing but connects me more fully to the ground, which is to be connected to the reality of one's body. We are creatures of the earth, enlivened by the spirit of the universe. Our humanity depends on this connection to the earth. When we lose this connection we become destructive. We lose sight of our identity with other people and other creatures, since we deny our common origin. We withdraw into our heads, into a world of our own creation where we see ourselves as special, omnipotent and immortal. The more we withdraw upward from the ground the larger our self-image grows. In this airy world there are no feelings of sadness or joy, of pain or of glory. There are no real feelings, only sentimentality.

Like so many other modern individuals I have been too egotistic, too narcissistic. I have needed to come down from my superior position, which I constructed in order to deny the humiliation I was made to feel as a child. Perched on this elevated platform, I was frightened to fall or to fail, for my identity was tied to my superiority. Fortunately I retained some identification with my body, which made me realize that any joy I hoped to find would be in the realm of the body with its sexuality. Coming down to earth for me was a long and difficult process, but when I did feel my feet connected to the ground, it was an experience of joy.

I am more in touch with my body that I have ever been, more

aware of its tensions and more conscious of its weaknesses. By the same token, I can sense my feelings more easily. Thus my anger will arise more quickly when I am provoked or hurt but I can also express it more appropriately. The result is that I am less frightened or anxious than I have ever been. If one is not frightened, one can accept life as it comes. This gives me a sense of inner peace which is the basis for joy. And I often have a sense of joy, which I experience in connection with the natural beauty in the people and things that surround me.

When one lives in terms of survival, meaning is attached to behavior and to objects that further survival, such as being good, being strong and having power. Since it is the nature of the human mind to look for meaning, individuals who are oriented to joy find meaning in attitudes and behavior that promote joy. Thus, I attach meaning to such attitudes as dignity, truthfulness and sensitivity. I aim to act in such a way that I can feel proud of myself, and to avoid any action that can cause me to feel ashamed or guilty. Dignity stems from the feeling that I can hold my head up and look someone straight in the eye. Truthfulness is a virtue but it is also an expression of respect for one's own integrity. When one tells a lie, the personality is split. The body knows the truth that the spoken words deny. This split is a very painful condition and is justified only when telling the truth poses a serious threat to life or integrity. Many people lie without feeling any pain, but this merely denotes that they are out of touch with their bodies and are insensitive to their feelings.

Sensitivity is the quality of a person who is fully alive. When we deaden ourselves, we lose our sensitivity. Thus children are the most sensitive individuals, as we all know. We need to be sensitive to others, but also to ourselves. If we are not sensitive to ourselves, we cannot be sensitive to others. The problem is that an insensitive person is unaware of his lack of sensitivity. I am not speaking of alertness, which is a state of heightened tension. Sensitivity is the ability to appreciate the fine nuances of expression associated with life, both human and non-human. Such sensitivity depends upon an inner peacefulness stemming from a lack of struggle, or of effort. These are the values that give true meaning to life, for they are the qualities that promote joy.

CHAPTER 12

PASSION AND THE SPIRIT

The Surrender to God

The dictum that man does not live by bread alone is well known, but it is not taken seriously in this culture, which is preoccupied with material things. To understand this preoccupation we must recognize that it stems from an identification with the ego and its values. The ego values objects and activities which serve to enhance an individual's image in the eyes of others. The accumulation of property serves this purpose, as does money, power, success, fame and position. Since the ego is an integral part of the human personality we are all interested in our image and our status in the community. A serious problem arises when the pursuit of ego values becomes the dominant activity of a culture. The result is that other, more important and deeper values, which we call spiritual, are ignored or devalued because we do not see their relevance to our daily lives. The opposition of materialism and spiritualism allows no reconciliation since they are either-or-concepts. If we use the term ego values to characterize the pursuit of material things, then the enhancement of spiritual feelings belongs to the realm of body values. The ego and the body merely reflect two different facets of the human personality. Both are essential to the healthy functioning of an individual.

Body values are supported by any object or activity which furthers the good feelings of the body, and include love, beauty, truth, freedom and dignity. These are inner values related to one's sense of self, as opposed to the ego, or material, values which stem from one's relationship with the outer world, and with the outer aspects of one's being. Inner values are true spiritual values since they are associated with activities of the spirit and give rise to strong feelings or passions. On the other hand, no one is really passionate about ego values, although many men are driven by an intense ambition to achieve them. The drive or ambition to become famous or the obsession to be rich does not arouse good body feelings. One might say that it feels good to be rich, but that feeling is related to the ego's perception that wealth provides security and power. To a primitive person the idea of wealth would not arouse much feeling, whereas dignity, honor and respect would evoke strong positive feelings. The lack of identification with these values lies at the base of the social troubles that plague our societies today.

Another spiritual value which is largely absent from our culture is the feeling of identification and harmony with nature, with one's environment and with the members of one's community. The primitive person is very closely connected emotionally with his environment since he is fully dependent on it for his survival. The modern individual, who in fact is equally dependent on the natural environment for his survival, has become alienated and dissociated from the natural world by his identification with his ego. Thus while he believes that he is more secure than the primitive who uses magic to increase his sense of security, modern man is deeply insecure on a body level because of the loss of his connectedness to the self, the earth and the universe. All religious activity aims at promoting these inner, spiritual or body values. They reflect the good feelings that stem from a sense of harmony and connectedness with the forces in nature and in the universe. If we substitute the word "God" for these forces, we can appreciate the power of religious feeling. When these feelings are strong, they constitute a passion which excites the spirit and maintains it at a high level of charge. When that passion or any aspect of it, like a passion for beauty, is present in an individual, I believe that it is impossible for him to become depressed, anxious or compulsive. In this time when spiritual or inner values have been lost, when religion has lost its power to influence feeling and behavior, depression and emotional distress have become endemic. On the other hand, I doubt that a belief system, religious or otherwise, can substitute for the feeling of passion. A feeling of passion can develop in an individual when he surrenders his ego-controls, freeing the body from its bondage to the will and to ego values. This surrender is the basis of religious healing, in which the surrender is to God.

The problem with some practices of religious healing is that the surrender is not to God but to a representative of God, or to a doctrinaire order which demands submission to an authority. This is similar to what happens in cults, where there is also a surrender of the ego to the leader with a resulting sense of freedom and the feeling of passion. Submission is not a true surrender and the spirit will rebel sooner or later against the loss of freedom to be true to one's self. I believe that true healing has to come from within the individual and not from an outside force. God plays a role in self-healing, for the healing force is the spirit of God that is within the body. That spirit is, of course, the spirit of the individual—the vital force that maintains his life, moves his body and creates the feeling of joy. But as we saw in the earlier chapters, the surrender to the body evokes a fear of death, a fear that one would not survive if one gave up ego-controls. The patient has no faith because the faith he had as a little child in the love of his parents was betrayed and he felt that he would or could die. But though surrender is frightening, it is the only way one can heal the wounds of childhood. It takes faith to let go or give in to the body, to the darkness of the unconscious, to the underworld of our being. It also requires a guide, a person in whom one can have faith because he has traversed the unknown in his own healing process and in his search to find the God within his being. At the same time that one connects with the God within, one also connects with the God without, with the cosmic processes that brought life into being and upon which our lives depend. Although we moderns are vastly more knowledgeable than primitive man, we have the same need for harmony in our relationship to nature and the universe.

Before we lost our innocence and became self-conscious we sensed this harmony. Some of us may recall that we felt this connectedness and harmony when as small children we experienced joy. When my son was about five years old, I made an effort to get him to go to Sunday School. My argument was that he would learn about God. He said, "I know about God." When I asked him what he knew, he pointed to some flowers that were growing in the garden near where he was standing and replied, "He is there." I sensed that he had a feeling about God which was more important than what he could learn in school and I abandoned my effort to get him to go to Sunday School. I felt sure that if he was aware that God was in the flowers, he also knew that God was in his own body too. This belief that all living things have a godlike quality is one of the principal concepts of the Hindu religion, which postulates that the Brahma essence is an attribute of all creatures. Primitive man believed that there was a spirit in all things, living and nonliving, which must be respected. Rivers, lakes, mountains and woods and all the things in them were animated by a spirit just as man was.

Animism, as this belief was called, was the first religious system. Given that small children think as primitive people do, it is not surprising that my son would spontaneously see God in all living things.

In early prehistoric days man lived fully in the natural world as one animal among many. It was an age of innocence and also of freedom. In mythology it was a paradisiacal age because eyes were bright and hearts were joyful. There was also pain and sorrow, for those feelings cannot be separated from pleasure and joy any more than night can be separated from day, or death from life. But a life in which there is pleasure and joy can make pain and sorrow bearable. Such a life contrasts sharply with modern existence in which there are few real pleasures and little or no joy. One has to be blind not to see this reality in the faces and bodies of people one meets on the streets or in other public places. Mostly, faces are tight and drawn, jaws are grim, eyes are dull or frightened or cold. This is evident despite the masks people wear to hide their pain and sadness. Bodies are frozen or disjointed, terribly overweight or too thin, rigid or collapsed. There are many exceptions to this description, but real beauty is rare and true gracefulness almost nonexistent. It is a tragic scene. In contrast, I saw a picture of a girl in a television documentary of one of the poorest tribes of people. They were nomads living in the Sahara Desert. The girl was carrying on her back a bundle of wood which she had gathered and was bringing to the camp for the evening fire. Since nights are bitterly cold in the Sahara, that bundle of wood was her contribution to her people. It was an expression of her love and her body reflected the joy she felt. Her eyes were shining and her face was radiant. I have never forgotten that picture.

I have not seen such a face in many years, but I remember seeing such faces in young women when I was a boy in New

York City. It was another time period and, I might say, another world. There were no automobiles and no refrigerators. Ice was delivered by an iceman and coal by a horse-drawn wagon. It was a slower time and a quieter time. People had time to sit on the steps in front of their houses and talk to one another. It was far from being paradise, and I was not a happy child, but I do recall joyful periods when we children played our games on the streets. Compared to that time, the city of New York, where I still have my office, has an unreal and almost nightmarish feeling.

Older people generally speak of the past in more favorable terms than the present. That was true even when I was young. It may be attributed to the fact that one saw his past with youthful eyes, with more excitement and more hope. But while that may be true, it is equally true that the quality of life has greatly deteriorated over my lifetime. While I feel more joy now than I have ever known, I believe that in every large city there has been a progressive loss of those qualities that contribute to the joy of life that is directly proportional to the increase of wealth and power. We have become a materialistic culture dominated by economic activity aimed solely at the increase of power and the production of things. The focus upon power and things which belong to the outer world undermines the values of the inner world—values such as dignity, beauty and grace.

I believe that the loss of moral and spiritual values is directly related to an increase of wealth. It is said that a camel could sooner pass through the eye of a needle than a rich man enter the kingdom of heaven. But that kingdom is the kingdom of God on earth where joy is possible. Unfortunately, man was expelled from this kingdom—which was the Garden of Eden—for having disobeyed God's injunction against eating the forbidden fruit of the tree of knowledge. But having gained knowledge, he became homo sapiens, which removed him from the purely animal state to the human condition. That move, the first small step toward man becoming a civilized creature, took a long time. Succeeding steps happened more quickly. From the Stone Age to the Bronze Age was a matter of four to five thousand years; from the Bronze age to the Iron Age took less than two thousand years. The pace of civilization quickened as man's knowledge grew, and with that growth there was a corresponding development in his conception of the godhead. The idea of an all-powerful male God, God the father, developed relatively recently and is limited to the religions of Western civilization. In the earliest religion, animism, all the spirits of nature were worshiped. Polytheism represented the worship of male and female gods and goddesses, each associated with specific aspects of human life. The rise to supremacy of a single male god was associated with the rise to power of a male ruler, the all-powerful king who was seen as a descendant or representative of the god. The god or gods no longer resided in the earth. They first moved to a mountain top—Mt. Olympus, where the Greek gods lived —and then the supreme God was removed to some remote place in heaven inaccessible to mortal man.

This process of separation of the divine from the secular represented a progressive demystification of nature and the body. The earth was seen as a mass of matter which, when activated by the energy of the sun, could produce plants. Man then learned how to control this natural phenomenon through agriculture, which provided him with a reliable source of food. Then, with the introduction of machines and chemical fertilizers, his power to do this seemed unlimited. We all know this story. But we have also become aware that there is a danger in this process. We are learning that we interfere with nature's own ecological balance at our peril. But we have done the same thing with our bodies, reducing them to biochemical processes and thereby robbing them of their godlike nature. Modern man in Western culture has lost his soul, as Jung pointed out.[1]

One could argue that the growth of civilization has been man's greatest achievement, his crowning glory. I would both agree and dissent. Civilization is identified with the life of the cities, but if today's large cities are the glory of man, they are also his shame. Few are free from the pollution of their air, the hyperactivity, the traffic, the noise, the violence and the dirt. There are always a few corners of quiet beauty but they are overwhelmed by the ugliness of modern advertising, expressing its obsession with material goods and sex.

Demystification removes an object or a process from the realm of the sacred to that of the vulgar. The sacred object becomes a thing, the sacred process becomes a mechanical operation. That has become the fate of the human body and its sexuality in the twentieth century. The sexual act, which is the communion of two individuals engaged in the sacred dance of life, has become for many people a performance and an ego trip. We may need, for special purposes, to see the bodily functions objectively as biochemical or mechanical processes, but we must not lose sight that there is a deeper reality to all living processes. Love can never be explained biochemically or mechanically any more than the power of the words "I love you" to arouse feeling can be explained by the auditory waves that carry the sound. Love is a state of intense positive excitement in the body, but that tells us little more than that life itself is a state of excitement. I would characterize love as the ultimate expression of life, because as the force behind the reproductive function, it is the creator of life. Reducing life, love and sex to physiological processes ignores the

[1] Jung, Carl, *Modern Man in Search of a Soul* (New York: Harcourt Brace Jovanovich, 1933).

emotional side of the body—activities which make them expressions of the body's spirit.

Eastern philosophy and religion do not separate or dissociate God from nature or the spirit from the body. The Chinese believe that all processes in nature and in the cosmos are governed by the interaction of two principles or forces, Yin and Yang, which when they are in balance guarantee the wellbeing of the individual. Hindu thinking recognizes an energetic force called Pranha, which is the breath. Bioenergetic Analysis uses an energetic principle to understand living processes and works with an energetic concept using breathing to free an individual from the tensions in his body that bind his spirit and limit his freedom. Eastern thinking is rooted in the view that man is not master of his life, that he is subject to forces which he cannot control—forces which can be subsumed under the terms "fate" or "karma." In contrast, Western scientific thinking sees no limits on man's possible power to control life. This view is based on our identification with the mind and its imaginative processes, which are not limited in time, space or possibilities of action. In contrast, an identification with the body forces man to realize the limitations of his being and the relative impotence of his actions.

The Eastern attitude to life has been described as fatalistic. Man is seen as powerless to change the course of events. Common sense would, therefore, advise acceptance and surrender. Such an attitude is rejected by most Westerners who would regard it as defeatist. Western man is encouraged to fight, to struggle, to believe that where there is a will, there is a way. The will is a very valuable function in life when it is used properly. Its place, however, is in emergency situations where a tremendous effort must be made in the interest of survival. To stay in control and not panic is a function of ego-control operating through the will. Losing one's head in a situation of danger is life-threatening. To attack a threatening enemy requires will because the body's tendency is to escape. Seen in this light the will is a positive force. But it has no place and becomes a negative force in situations in which there is no danger and activity should be pleasurable. Imagine using the will to enjoy a sexual relationship! As I pointed out in this book, joy is dependent on a surrender of the will and of the ego.

This surrender of the ego allows the person to turn inward, to hear the voice of God. Meditation, as it is practiced in the Eastern religions, is a means whereby the individual can shut out the noise of the external world so that he can hear his inner voice—the voice of the God within him. To close out the noise of the external world, one needs to shut off the flow of thoughts, which is called the stream of consciousness. This stream of consciousness arises from the constant stimulation of the forebrain from subliminal muscular tension. It ceases when one goes into a state of deep body relaxation in which breathing is full and deep. One has, in effect, let go of unconscious control associated with an inner state of alertness. When this is done a sense of inner peace pervades the body. Consciousness is not dimmed. One is fully aware but the awareness if not focused. One is not unconsciously poised to meet a danger.

I have been in this state and it is a beautiful experience. It approaches the feeling of joy. One might say it is a low-keyed feeling of joy. I had such an experience following a week in which I was literally floored by my doctor treating an attack of sciatica. Persistent pain in my lower back, buttocks, and right leg, with paraesthesias indicating some nerve involvement, persisted for several months despite treatment. I called a colleague who was an orthopedic physician and familiar with Bioenergetic Analysis, who advised me to lie on the floor with my knees bent and my feet resting on a box of books. I was to eat lying on the floor, to sleep on the floor and to read on the floor. He recommended a sort of crawling movement if I needed to go to the bathroom. This position on the floor took the weight off my lower back, allowing the tense muscles to relax. But its effect on my personality was unexpected. It quieted me down, down, down. On the fifth day I sat outside on a chair in the sun, my hands on my lap. I wasn't thinking. I could feel the deep inner pulsation of my body as I breathed deeply without any conscious effort. I didn't meditate. I just sat like a cat regarding my surroundings. It was heavenly.

My sciatic condition wasn't cleared up by the week on the floor, although the pain was lessened. Perhaps I should have stayed longer, but I had things to do and I was scheduled to leave for Greece ten days later. In Greece I had a massage and several acupuncture treatments which helped a little. My condition was improving but I could still feel the pain. One day I realized I was completely free of pain, and had been for some days. As I tried to think back to when the pain stopped, I could only remember an incident which had happened at about the time the pain disappeared: I had become very angry with a colleague who was associated with the stress which I knew had been the cause of my sciatic condition. As I spoke to him a feeling of anger flared through my body in a wave of excitation that discharged all the tension in my back and released me from pain. It made me realize that anger when properly expressed is a healing force.

This anger was the voice of the God within me. It was not something I did in the sense that it was a conscious and deliberate action. It just happened. Some force inside my body erupted as a surge of anger. On another occasion I experienced a surge of love which transformed me. In fact, every emotion—fear, sadness, anger, love—is a pulse of life, a surge of feeling from the core of one's being. This core is constantly pulsing, constantly sending out impulses which maintain the life process. It is the energetic center of the organism, as the sun is the energetic center of the solar system. It is responsible for the beating of the heart, the rhythmic in-and-out of breathing, the peristaltic activity of the intestines and other tube-like structures. Hindu thinking recognizes energetic centers called chakras, but I believe that there must be one primordial or principal center to maintain the integrity of such a complex organism as a mammal. Great religious mystics have placed this center in the heart, which they see as the abode of God in man. It is certainly the seat of the impulse of love, which is the wellspring of life and the source of joy.[2] Although we are familiar with the pulsation of the heart, the fact is that every cell, every tissue and the whole body pulsates, which means that it expands and contracts rhythmically. The heart expands and contracts as it beats, the lungs expand and contract as we breathe. When that rhythmic pulsation is free and full we feel pleasure. We are pleasurably excited. When the excitement mounts so that the pulsation is more intense, we feel joy. If the intensity of the excitement reaches its maximum or acme, we experience ecstasy. In the absence of any excitement or pulsation, the organism is dead. Excitation is the result of an energetic process in the body related to metabolism. A source of energy—food—is metabolized or burned to release energy necessary for the living process. If life is seen as a fire continually burning in a water medium, love can be described as its flame. Poets and songwriters have used this metaphor for ages. But it is more than a metaphor. A person in love literally glows, the flame of his feeling shining in his eyes. Such intensity of feeling or of excitation can be described as passion.

Love, passion, joy and ecstasy are terms also used to describe man's relation to God, the god within and the god without. There is a fire in the universe and an energetic pulsation related

---

[2] See Lowen, Alexander, *Love, Sex and Your Heart*, for a fuller discussion of these concepts.

to a process of expansion and contraction. Since our life stems from and is part of that process, we feel identified with it. Some mystics can actually feel the connection between the beat in their heart and the pulse of the universe. I have actually felt my own heart beat in rhythm with the hearts of birds. In a city they are the only creatures who are really free.

The phenomenon of empathy in which one can sense another person's feelings happens when two bodies vibrate on the same wavelength. Empathy is the basic tool of the therapist. It is absent in people whose bodies are too rigid or frozen, so that there is little pulsatory activity. When one's body is more alive, a person is more sensitive to others and to their feelings. Of course, when one is more alive one is more capable of love, and of joy.

While love is the source of life, it is not the protector of life. It is naive to believe that being a loving person will assure that one will not be hurt in life. All individuals start life loving and lovable, which doesn't prevent the attacks and traumas to which so many are subjected as children. The pages of this book testify to the pain and damage they suffer. A living organism would not survive long if it had no means of defense. In most organisms that defense takes the form of anger. We normally respond with anger to an attack on our integrity or our freedom. Anger is one aspect of life's passion. A passionate individual will passionately defend the right of every individual to life, liberty and the pursuit of happiness. A just God would have it no other way.

The Dancing Spirit

Joy is an extraordinary feeling for adults whose lives revolve around ordinary activities and ordinary things. These things and activities can give us pleasure, but the excitement associated with them rarely reaches the height of joy. The main reason for the lack of joy in ordinary activities is that they are ego-directed and controlled. Small children can experience joy easily in ordinary activities because none of their simple actions are ego-controlled. A child acts spontaneously, without thinking or planning, in response to the natural impulses in his body. In contrast to adults whose movements are largely directed and controlled by the ego, a child is moved by feelings or forces which are independent of his conscious mind. The difference between moving from the ego or a conscious center and being moved by a force emanating from some deep center in the body distinguishes the extraordinary from the ordinary, the sacred from the secular, joy from pleasure. When I saw my young son jump for joy, I realized that he didn't jump in a conscious or deliberate way, but was lifted off the ground by a surge of positive excitement which propelled him upward. He had a "moving" experience. All extraordinary experiences have the quality of being "moving" experiences. That quality also attaches to most deeply religious experiences, which a religious person would see as manifestations of God's presence or grace. This is a valid interpretation since the force that moves the person has to be greater than the person's conscious self.

Deeply moving experiences occur in situations that have no direct connection with religion or the concept of God. The most common of these deeply moving experiences, which has no religious connotation for most people, is falling in love. And what a joyful experience it is to be in love! That happens when our heart is touched or moved by another individual. Heartfelt love for any creature or individual can also be regarded as a manifestation of God's grace. In surrendering to love we surrender to the God within us. Love moves an individual to closeness with the love object, aiming at a physical closeness or contact with the loved one and, in sexuality, an energetic fusion of the two organisms. The feeling that brings two individuals together in love is passion, which also describes the desire for closeness with God. Passion denotes an intensity of feeling which moves an individual to transcend the boundaries of the self or the ego. When that happens in a sexual orgasm which embraces the whole body in its convulsive movements, it is the experience of transcendence par excellence. It doesn't happen much in our culture because sex and sexuality have been removed from the realm of the sacred to that of the ordinary and secular. Sex is something one does to relax or to relieve a tension, not an expression of passion.

One other activity which partakes of this quality of being a "moving" experience, although to a much smaller degree than sex, is dancing. Normally we are moved to dance by music. When we hear dance music our feet and legs cannot keep still. If the rhythm is strong and persistent we can be swept up into it and carried away. Such dancing is a moving experience which can lead to a transcendental state. For most primitive people dancing is a part of their religious ceremonies. But whether associated with religion or romance, dancing always leads to joy, and many times also to love. The key to the transcendence of the self is the surrender of the ego.

All religions proclaim that the surrender to God is the way to joy. Sri Daya Mata, the spiritual head of Self-Realization Fellowship, an organization founded by Paramahansa Yogananda, a famous Indian guru, says, "No human experience can match the perfect love and bliss that inundate the consciousness when we truly surrender to God." While his statement represents a basic Hindu philosophy, it echoes similar ideas which can be found in all religions. I, too, believe that is the true way. However, people have lost their way to God, otherwise it would not be necessary to guide them or to counsel them. Little children can experience joy without need of guidance or counsel, which must mean that they are in touch with the God within them. For the adults who have lost touch with the God in them, recovering that contact is not an easy task. Sri Daya Mata gives some sound advice on how to do this, but the best advice is rarely effective because one can't follow it. One is blocked by unconscious fears which make surrender a dangerous undertaking, as we have seen in the cases discussed in these chapters.

Eastern religion offers procedures which are helpful in promoting the surrender to God. The best known of these practices is meditation, a procedure which allows the individual to turn inward to make contact with the God within. The chanting of a mantra or the making of a sound helps shut out the noise of the external world, quieting mental activity. Meditation is now widely used in the West as a relaxation technique, a means to reduce the enormous stress to which so many individuals in the industrialized world are subject. To achieve the surrender to the inner God meditation must be carried on for an extended time. Most monks who strive for this deep contact retire from the world for long periods and forsake all worldly pleasures. Withdrawal from the external world is also a feature in the Christian religion for those persons who wish to live a deeply religious life undisturbed by the cares and concerns of the external world. Praying, singing and contemplation are the activities which, for Christians, promote contact with the God within. Many individuals in the West make these practices a part of their daily life just as people in the East use meditation for the same purpose. But as the pressure and pace of life increase with the growth of commerce and technology, the religious life in both the East and the West seems to disappear more and more. That disappearance coincides with the loss of contact with nature, with the body and with the spiritual aspect of life.

But is it necessary to withdraw from the world to be spiritual and experience a contact with God? That cannot be a practical or realistic way of life for most people who are involved with the ordinary activities of making a living and raising a family. However, when these activities are undertaken with a spirit of reverence for the great forces in nature and the universe which make life possible, the ordinary activities of life have a spiritual quality. Spirituality is not a way of acting or of thinking; it is the life of the spirit which is expressed in the spontaneous and involuntary movements of the body in actions which are not ego-directed or controlled. These movements are pulsatory and rhythmic like the beating of the heart, the peristaltic action of the intestines and the respiratory waves that flow upward and downward through the body. The natural vibratory activity of the body which underlies the above functions is, in my opinion, the basic manifestation of the living spirit. When this vibratory activity ceases we become aware that the body is dead, that the spirit is extinguished and that the soul has left the corpse. When a person's eyes sparkle, it denotes a highly charged vibratory activity in the eyes which also produces a radiation. Vibration is also evident in the voice. Here, too, a dead voice denotes a loss or diminution of one's aliveness or one's spirit. This involuntary activity in the body is what we perceive as feeling. Only living creatures have feelings because feeling is how one experiences the life of the spirit. When one's spirit is low, feeling is low. High spirits are reflected in strong feelings. It is the spirit in us that moves us to love, to tears, to dance and to sing. It is the spirit in man that cries out for justice, fights for freedom and rejoices in the beauty of all nature. It is also the spirit that moves us to anger. The strength of a person's spirit is reflected in the intensity of his feelings. When the spirit is strong the person has a passionate nature. In such individuals, the flame of life burns bright and the individual senses that his spirit reflects God's love.

Spirit is not a mystical concept. The spirit of a person is manifested in his aliveness, in the brightness of his eyes, in the resonance of his voice and in the ease and gracefulness of his movements. These qualities are related to and stem from a high level of energy in the body. This is not understood in a machine-driven culture that equates energy with drive and with the power to do. The energy of life acts differently. It functions simply to protect and promote the well-being of the organism and to perpetuate the species. The well-being of the organism is experienced in the good feelings of the individual, which range from pleasure through joy and reach heights of ecstasy at times. These good feelings reflect the degree of positive excitation in the body and are manifested in the body's pulsatory activity. When the pulsation is strong and deep, it is also generally quiet and contained, which is seen in the quiet beating of the heart and in deep and easy respiratory activity. This steady, rhythmic activity is experienced as pleasure. The moment the individual pushes to achieve a goal, the body is under pressure and the easy, steady rhythm of pleasurable movement is lost.

Pushing and driving develop when a person senses the need to mobilize extra energy for a task. That mobilization requires the use of the will, which creates a stress on the organism. Individuals with a high energy level are relatively free from stress in their normal activities. Their bodies are more relaxed, their movements are more graceful, and their behavior follows a quieter pattern. Like a high-powered car, they can climb a hill with less strain. Individuals whose energy is low have to push themselves, which drains energy because of the stress, leaving them tired and feeling that they can't succeed or get through unless they make an even greater effort. They are often afraid to slow down or stop, for fear that they would fail or that they might not be able to start again. Many keep going to avoid becoming depressed. The most common complaints of people in the industrialized world are tiredness and depression.

Anyone familiar with modern life knows that the pace of activity has increased enormously in this century in proportion to the increase in the speed of travel and communication. How can one surrender when one is going so fast that one can't stop? How can one sense the God within when one is going at 60 or more miles per hour on a crowded highway? Yet in this hectic and driven culture some people pride themselves on being in the fast track. The faster they move and the more they do, the less time they have to feel, which may be one reason they keep so busy.

The pulsatory activity of life is clearly seen in the animal called the Medusa, or jellyfish. Pulsation creates internal waves in the jellyfish which move it through the water. The same pulsatory activity can be seen in worms or snakes, in the form of waves which move these creatures through space. In the higher animals the pulsatory activity is more internal as in the peristaltic waves that move the food through the intestines. Since the heart is the organ in the body that pulsates the strongest, it is regarded by many mystics as the abode of God. We may wonder, however; is God the force that creates the pulsation—or is he the pulsation itself? Sensing this spontaneous pulsatory activity of the body one can believe that it is a direct manifestation of the spirit within. There is also pulsatory activity in the heavens, in the whirling of the heavenly bodies, in the periodic emission of light and radio waves. Sensing the harmony between the internal pulsation of our body and that in the universe, we feel identified with the universal, with God. We are like two tuning forks vibrating to the same pitch.

Since pulsation is an aspect of the natural world, man could well believe that there is a sacred spirit in all things. This belief is the basis of animistic religion. With increasing knowledge, objectivity and power, man's ego denied a divine spirit to nature and to other creatures, seeing himself as the sole being who partakes of the Godhead. Some individuals have actually reached the point where they deny any connection with the divine or the God within. One can only reach this conclusion if one has lost all contact with the pulsatory activity of his body. For such a person, the heart beats because it receives signals from the brain which has been programmed genetically to send these signals, similar to the way that a computer can run a system once it has been programmed. That our brains are programmed by heredity and experience to coordinate the complex computer operations of the body cannot be doubted, but that leaves open the question of who programmed man. The religious answer is God, who created man, which implies the existence of an active God force by which to explain evolution. A mechanistic view of life leaves no room for a divine spirit and no possibility, therefore, for a moving experience which gives meaning to life. If we recognize that the living spirit within an organism is godlike we can avoid the conflict between a mystical, religious view of life and a mechanistic one.

The denial of the spirit characterizes the narcissistic individual of our time.[3] The narcissist sees the world and life in mechanistic terms: stimulus and response, action and reaction, cause and effect. There is no room for feeling in this character structure. Feelings are imprecise, immeasurable, often unpredictable and certainly not rational. In the narcissist the life of the spirit is unknown and denied. He exists consciously in his head, is dissociated from his body and lives the life of his mind. Narcissism is alien to children whose lives revolve around the fulfillment of desire, the joy of freedom and the pleasures of self-expression. Children like to be admired, as we all do, but they will not sacrifice their feelings to be special or superior. Children compete and want to be on top because they are very self-centered. They are passionate creatures who want it all, but they are not egotistic. They love and want to be loved because their hearts are open. As two parents remarked about their nine-month-old daughter, "She is a bundle of joy." This is what childhood is about. Children feel the joy of life when they are loved and bring that joy

[3] Lowen, Alexander, *Narcissism, The Denial of the True Self* (New York: Macmillan Publishing Company, 1985).

to others. They are the innocents and the powerless ones and so are vulnerable to the negativity and hostility of the adults about them, including their parents. People who have lost their joy cannot stand to see others have it.

We have seen in these pages how the innocence of children is destroyed and their freedom lost. A harried parent cannot stand the crying of a baby and threatens it. A frustrated parent cannot allow the child to have the joy which he or she cannot feel and punishes it. A rigid parent cannot tolerate the exuberance and spontaneity of young life and destroys it. Not all children survive the insensitivity and cruelty of their caretakers. Child abuse has resulted in the deaths of many. Most parents are ambivalent. They love the child but they also hate it. I have seen a mother look at her daughter in my office with such black, hate-filled eyes that I was horrified. However there is also some love. Children do not understand ambivalence, which is a sophisticated concept beyond their comprehension. When they feel the hatred, they cannot sense or believe in the love. When they feel the love, they forget the hatred. They will learn about ambivalence and, in turn, become ambivalent themselves.

When a young child senses the hate and violence in a parent, it cannot help but think that its life is threatened. The experience of that threat is a shock from which the organism may never fully recover. The child is actually threatened in two ways: one is the possibility of violence—that it will literally be killed, which sends a wave of terror through the child's body. On the body level that memory will never be fully erased. The other threat is rejection and abandonment which, for a child, is also a death threat. These threats are not carried out but a very young child cannot imagine that they are intended only as a scare. It must submit, it must curb its aggression, it must dampen its excitement and to do this it must restrict its breathing.

Bioenergetic Analysis aims to help a person breathe more

deeply because without deep breathing, one doesn't have the energy to feel the passion of life. However, getting patients to breathe deeply is a difficult task. Breathing is an aggressive action. One sucks air into the lungs. Unfortunately, most infants are discouraged from being aggressive. Many are handicapped from birth by being denied the emotionally fulfilling experience of being breastfed. They are given a bottle, which places them in a passive position since it doesn't take much sucking to get the milk. Breastfed babies can suck energetically and, as a result, their breathing is more energetic. On the other hand I have found that breastfed babies can be severely traumatized when they are weaned too early. In my view, normal breastfeeding should last three years, as it does in primitive societies. That is very rare in our culture because women are too pressured to be able to give that much time to an infant. Many have to go back to work shortly after the baby is born to help support the family. One sees this lack of fulfillment in patients whose breathing is shallow and who complain of feeling empty, insecure and depressed.

But the lack of adequate breastfeeding is not the only cause for the sadness and despair that afflict so many people. The child's need for loving contact with the mother cannot be met by mothers who are themselves unfulfilled individuals, whose bodies do not give off a strong positive excitement which would stimulate and excite the child's body. Mothers are stressed by babies who demand more contact and attention, and babies are stressed by mothers who cannot respond to those demands. In the conflict that develops between them, the baby feels threatened in its existence. Survival demands adaptation which means that the child learns to function on a lower energy level with a reduced respiratory function. Getting such patients to breathe deeply generally brings up a fear of dying. I have had several patients complain that as they breathed more deeply they felt a blackness come into their head, and had the feeling that they would faint. It was as if they felt they were going to die—a very scary experience. However, it is an irrational fear. One doesn't die from breathing deeply. One might hyperventilate and faint, but there is no danger in that. And even that won't happen if one can keep breathing through the fear. It is the stopping of breathing that cuts the blood off from the brain, creating a sensation of darkness and ending in a sudden faintness. I advise my patients, therefore, to remain focused on their breathing. One patient, a very frightened woman, found the courage to stay with the breathing and to her amazement, the light in her head didn't dim and she didn't faint. She was very excited by this result and kept exclaiming, "I went through it! I went through it!" She left the session in a state of euphoria.

I am convinced that we all have to face our fear of death if we wish to enter the kingdom of heaven that is within us. The angel with the flaming sword that guards the entrance to the Garden of Eden, the original paradise, is also within us. It is the parent with the cold, hate-filled eyes who could have destroyed us for disobeying. It is the guilt that says, "You sinned. You have no right to happiness." And, finally, it is your anger turned inward and against yourself because of guilt, shame and fear. This exhilarating experience of the patient who "went through it" does not guarantee that she is free from the fear of death. Actually hers was a first step into the valley of death, a step she took without panic. There will be many more sessions in which she will face her fear of death as she asserts her right to selfhood. Each assertion, each deep breath, strengthens the life force in her and supports her desire to go deeper. Life and death are opposite states of being, which means that when one is fully alive there is no fear of death. An individual's personal death does not exist other than as a future event. It is an idea, not a feeling. If there is any fear in us we can ascribe it to this future event. If there is no fear in the personality then death is not frightening. Brave men can die without fear. As the saying goes: A brave man dies only once; a coward dies a thousand deaths. When the current of life flows freely through the body, there can be no fear since fear is a state of contraction in the body.

The surrender to God removes the fear of death because it activates the current of life which had been constricted by the ego in its attempt to control fear and other feelings. But by this very means it promotes life and healing. I had two patients who were at death's door, one from septicemia, the other, during open heart surgery. Both told me that, sensing the possibility of death, they put their life in God's hands. Both recovered and both stated that they believed this act was the turning point in the illness. There is nothing mystical about this phenomenon. The surrendering of the ego removes the defenses that block the flow of life, which can only have a beneficial effect on the body. The surrender of the ego also involves the surrender of the will, including the will to live. Life is not an action one can will. The will to live is a defense against an underlying wish to die.[4] It represents the attempt to overcome one's fear of death but doesn't remove that fear. What keeps life going is not the will but a continuing state of positive excitement in the body which is expressed as a wish to live. That excitement is generated by the pulsatory activity of the body, which is God-given.

One morning I awoke with the sweetest feeling in my body. It was as if my whole body was sugar or honey. Experiencing this feeling I thought, "If you are true to yourself, you are not afraid of death." It was such a beautiful and unusual experience that I wondered what had produced it. I could not recall any dream that night. I thought back to the events of the preceding evening and remembered that I had watched the movie *Platoon*

---

[4] See Lowen, A., *Love, Sex and Your Heart*.

on video. It is the story of a group of American soldiers in the
Vietnam War. Some of the soldiers of the platoon kill some
Vietnamese civilians mercilessly. Several others are angry at this
behavior and a conflict develops among the men. The conflict
ends with the killing of two members of the platoon by their
own men. As I thought about the movie, I came to the conclusion
that the senseless violence of the soldiers was due to fear—not
just fear itself, but to the denial of fear. They were scared to
death, but instead of acknowledging their fear, they denied it
and killed others.

Fear is a natural emotion which all creatures share. For a
person to deny his fear is to deny his humanity. Feeling afraid
doesn't mean that one is a coward. One can act courageously in
the face of fear, which is true courage. When we deny fear, we
set ourselves above the natural world. Since the suppression of
feeling is done by deadening the body, the suppression of fear
operates to suppress anger, sadness and even love. We lose God's
grace and become monsters—that is, unreal. If someone pointed
a gun at me, I would be afraid that he might kill me. But the
fear of being killed is not the same as the fear of death. Since
death cannot be separated from life it is part of the natural order.
When it occurs as part of the natural order, we can accept it
with equanimity. When an individual is scared *of* death, it is
because he is scared *to* death. Thus, when a person is true to
himself, he is free from fear, including the fear of death. By the
same token, if we are not afraid of death, we can be true to
ourselves.

To be true to yourself means that one has the inner freedom
to sense and accept one's feelings and is able to express them. It
also means that one has no guilt about what one feels. When
there is guilt in an individual, he cannot express his feelings
openly and directly. He has a censor in his mind that monitors

all expression. This does not mean that one acts on all his feelings. We are not infants without egos. We know what behavior is acceptable to society and what is not. We have or should have a sense of self-possession which enables us to express or act on a feeling in a way that is appropriate and effective for our needs. Such conscious control is not based on fear. Fear paralyzes and one's actions become awkward and ineffective. One loses the lovely spontaneity which endows one's action with grace and graciousness. Self-possession is the mark of an individual whose statements and actions stem from a keen sensitivity to life and to others.

Joy is the experience of that lovely spontaneity which characterizes the behavior of children whose innocence has not been destroyed and whose freedom has not been lost. As we have seen, children lose their innocence and freedom rather early under the pressure of the harsh realities of modern family life. Survival, not joy, becomes the central theme of their lives. Survival demands sophistication, deceit, manipulation and a constant alertness based on fear. But survival is self-defeating, demanding a retreat from self-awareness, self-expression and self-possession. Life becomes a struggle, and even when one's situation as an adult poses no death threat, the average individual continues to struggle as if it did. Again and again, patients say to me, "I can't tell you what I think or feel. I'm afraid you'll reject me." One patient said, "I can't tell you that I love you. You will reject me." Another one said, "I can't show any anger toward you; you'll tell me to leave." Even saying this was a step toward freedom. To be open even with a therapist who is supportive of free expression takes considerable courage. That courage slowly but steadily grows in patients through the bioenergetic process; increasing the patient's energy, promoting self-expression and helping the patient understand his problem.

Therapy is not a matter of learning self-assertiveness. Such procedures encourage a pseudo-aggression which is a willed action, not a spontaneous one. Patients tell me, "Do you know what happened to me yesterday? My boss spoke to me in a condescending manner and without thinking I said to him, "Don't talk to me that way," and he apologized." The patient who told me this was even more surprised by her outspokenness than her boss. Having once broken the barrier of fear it becomes easier to open the door to freedom again. The initial breakthrough is a joyful experience which comes from the surge of life that flows through the body. One can have such experiences without undergoing therapy. A person who is told that she needs a biopsy to determine if a particular growth or lesion is cancerous will experience the same joy, the same sense of freedom from fear, when told that the biopsy proved negative. In this case, too, the joy stems from a surge of life. The difference between the two situations is that the therapeutic experience is not serendipitous. It is the logical outcome of a process of self-discovery. One feels more and more joy as one discovers more and more of the self. In a workshop recently, a participant turned to me with excitement, remarking, "It's the first time I felt my body doing it." What it was doing was pulsating. Her body had come alive as an independent force strong enough to overcome her sense that it was an object which her mind controlled. This happened because we had done considerable work deepening the breathing, using the voice and expressing feeling. These exercises were like priming a pump so that it could operate on its own.

When the body moves on its own in a total, organismic way, it is a moving experience. It is what happens when a child jumps for joy. One doesn't jump in a conscious sense; the body is lifted off the ground and the experience is felt as joyful. Some years ago I was walking down a country road in a very relaxed manner. I recall a step which produced an unexpected sensation. As my foot touched the ground, I sensed a flow through my body from the ground and felt two inches taller. I felt my body straighten and my head rise. It was a wonderful feeling. I couldn't say what caused it but it was associated with a surge of freedom and joy.

Freedom is the basis of joy. It is not just freedom from external restraints, although that is essential. More particularly, it is freedom from internal constraints. Those constraints stem from fear and are represented by chronic muscular tensions which inhibit spontaneity, restrict respiration and block self-expression. We are literally bound by these constraints. Every breakthrough represented by a surge of feeling is also a breakout to freedom. These breakthroughs and breakouts occur from time to time in the course of the therapy when a strong enough charge builds up behind an impulse to reach out, to open up, to express a feeling. I recall a session with Reich which had a liberating effect on me. The reader may recall that Reich's therapy involved the "letting-go" to one's breathing so that it would become deeper, freer and fuller. As I lay on the bed surrendering to my body, I felt myself rise slowly to a sitting position. The force in me which produced this action turned me around, and I got up on my feet. Without knowing what I was going to do, I turned, faced the bed and began to hit it with my fists. As I did this, I saw the face of my father on the sheets and I recalled that he had given me a spanking for coming home late one evening and upsetting my mother. I must have been about nine or ten. I had been out playing with my friends. I had completely forgotten the incident until it flashed into my mind as I was hitting the bed. Although this was not the first time I had had a spontaneous, moving experience in my therapy with Reich, I was awed by the experience and exhilarated at the same time. It was as if a hidden recess in my personality had opened up allowing me to step into a larger dimension of being.

The surrender to God is the surrender to the life process in the body, to feeling, to sexuality. The flow of excitation in the body creates sexual feelings when it flows downward and spiritual feelings when it flows upward. The action is pulsatory and cannot be any stronger in one direction than the other. I pointed out in my previous book that a person cannot be more spiritual than he is sexual nor more sexual than he is spiritual. Sexuality does not mean sexual intercourse any more than spirituality means going to church or belonging to a religious order. It refers to feelings of excitement in relation to a person of the opposite sex. Spirituality refers to feelings or excitement in relation to nature, to life and to the universe. The greatest surrender to God can occur in the act of sex if the climax is intense enough to send the person into orbit among the stars. In the total orgasm the spirit transcends the self to become one with the pulsating universe.

One of my former patients wrote me an account of an experience of surrender which illustrates the nature of the excitatory process. She described an evening with a male friend. They had spent the evening, at dinner and in his house, talking about his problems. He had been through a very difficult time some months earlier and she observed that he had not fully recovered. He looked like a zombie. She writes: "I was concerned about his persistent sad look and his slow movements. He didn't look healthy. He wasn't his old self. He didn't say much. We said good night and he went off to his room.

"In the morning he came to my room, which was unlike him, and asked if he could cuddle with me. The house was so cold we each had about six layers of clothing on, so I said, 'Okay.' He lay down with his back to me and I put my arms around him. He said, 'If you had a friend who was dead, what would you tell him to do?' I said, 'Well, if he's dead there is nothing I could say to him. But if he just felt dead, I would tell him to do his work, to get some help and to take care of himself, like I'm telling you.' He said he knew he was in trouble because his hands and feet felt frozen. He said that his recent experience had devastated him. He said he had spoken out about an unjust situation but it was very difficult because he was conflicted. As a child he had been abused for speaking out. He said, 'If only I hadn't been so abused,' he broke down and cried. I stroked him and told him that he needed to get help. He said that he had gone to some therapists, but when he felt rage, it had scared them. He shared a lot of stuff with me. Eventually he turned over to face me and he put his arms around me. I got a strong charge going through my body. It was vibrating. He said it was like having a big purring cat in his arms. The charge just kept building and then he too began to feel the charge. Our bodies just took off on their own, vibrating, pulsing and moving. At one point, I said, 'I'm not doing this; it's just happening.' And my God! We had this incredible full body orgasm and we were fully dressed. I was right with it. I didn't check out. I was fully present. My body did this incredible thing. It was wholly unexpected. I still have a nice hum and buzz from the experience. Now I understand what you mean by giving in to your body. It was a very enlivening experience.

"Well, I don't know what this will mean in the long run. I just try to live my life the best I can and enjoy what I can. This was one of my most interesting weekends."

This patient had spent many years in therapy and in working with herself. She was also professionally active in counseling, so that she had the background to understand what was happening and to go with it. She had developed a faith in life and a trust in her body which extended to God.

Sexual excitement sends a body spinning. We experience that

THE SPIRALING UNIVERSE

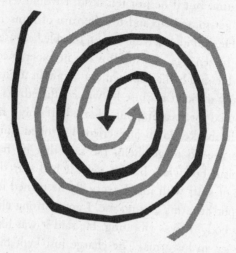

Schematic drawing of the spiral form G 10 of galaxy Messier 81
from a Mount Wilson photograph

(Reich, W., *Cosmic Superimposition*. Rangeley, Maine:
Wilhelm Reich Foundation, 1951, p. 61.)

FIGURE 6

spinning most visibly as we spin out of control in the convulsive
movements of the orgasm, which produce a feeling of ecstasy in
the individual. But an intense sexual excitement can actually
make one's head spin; this can be a joyful feeling if one is not
frightened of the sensation. Even a feeling of love can cause one
to circle around or encircle the loved person.

Reich had the brilliant conception that the energetic process
in sexual intercourse resembled the cosmic process he called superimposition. His theory was that when two energy systems are
attracted to each other, they begin to spin around each other as
they are drawn closer together. That process of cosmic super-

SPIRALING IN THE SEXUAL EMBRACE

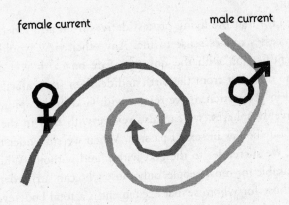

Schematic representation of the flow of
excitation in the sexual embrace of mammals

FIGURE 7

imposition can be seen in photographs of galaxies which show a spiral or whirling motion of stars as they spin in space. That motion is shown in a photograph of the spiral nebula known as G10 that appears in Reich's book *Cosmic Superimposition*. Reich saw the two arms of the nebula as energetic waves or currents which were drawing the stars of the nebula closer together as they spun around each other. (See Figure 6.) The active force that draws these stars toward each other is the power of gravitation, when objects in space get close enough they are drawn towards each other. In the animal world, we call the force that draws two individuals together love, or sexuality. Among mammals, where the male mounts the female in the sexual embrace, their posture and activity resemble the phenomenon of superimposition. The movement of the waves of excitation in the two individuals resembles the cosmic event just described, as shown in Figure 7.

The idea that the living process derives from and reflects cosmic processes makes sense to me. Any other view would deny our identification with the universe. Life on earth is a cosmic event no different from the birth and death of stars, albeit infinitesimal in proportion. If we rejoice with God in the spinning of the heavenly spheres, we can also rejoice with him in the spinning of our bodies in sexual passion. When we surrender to this passion, we surrender to the god within and without. While sex is pleasurable for most people, only those who can surrender their egos—those for whom sexual excitement is a total body event—can know the real joy of sex.

FOR THE BEST IN PAPERBACKS, LOOK FOR THE

In every corner of the world, on every subject under the sun, Penguin represents quality and variety—the very best in publishing today.

For complete information about books available from Penguin—including Puffins, Penguin Classics, and Arkana—and how to order them, write to us at the appropriate address below. Please note that for copyright reasons the selection of books varies from country to country.

In the United Kingdom: Please write to *Dept. JC, Penguin Books Ltd, FREEPOST, West Drayton, Middlesex UB7 0BR.*

If you have any difficulty in obtaining a title, please send your order with the correct money, plus ten percent for postage and packaging, to *P.O. Box No. 11, West Drayton, Middlesex UB7 0BR*

In the United States: Please write to *Consumer Sales, Penguin USA, P.O. Box 999, Dept. 17109, Bergenfield, New Jersey 07621-0120.* VISA and MasterCard holders call 1-800-253-6476 to order all Penguin titles

In Canada: Please write to *Penguin Books Canada Ltd, 10 Alcorn Avenue, Suite 300, Toronto, Ontario M4V 3B2*

In Australia: Please write to *Penguin Books Australia Ltd, P.O. Box 257, Ringwood, Victoria 3134*

In New Zealand: Please write to *Penguin Books (NZ) Ltd, Private Bag 102902, North Shore Mail Centre, Auckland 10*

In India: Please write to *Penguin Books India Pvt Ltd, 706 Eros Apartments, 56 Nehru Place, New Delhi 110 019*

In the Netherlands: Please write to *Penguin Books Netherlands bv, Postbus 3507, NL-1001 AH Amsterdam*

In Germany: Please write to *Penguin Books Deutschland GmbH, Metzlerstrasse 26, 60594 Frankfurt am Main*

In Spain: Please write to *Penguin Books S.A., Bravo Murillo 19, 1° B, 28015 Madrid*

In Italy: Please write to *Penguin Italia s.r.l., Via Felice Casati 20, I-20124 Milano*

In France: Please write to *Penguin France S.A., 17 rue Lejeune, F–31000 Toulouse*

In Japan: Please write to *Penguin Books Japan, Ishikiribashi Building, 2–5–4, Suido, Bunkyo-ku, Tokyo 112*

In Greece: Please write to *Penguin Hellas Ltd, Dimocritou 3, GR–106 71 Athens*

In South Africa: Please write to *Longman Penguin Southern Africa (Pty) Ltd, Private Bag X08, Bertsham 2013*